# THE FIELDS OF
# ELECTRONICS

# THE FIELDS OF ELECTRONICS

## Understanding Electronics Using Basic Physics

**Ralph Morrison**

A Wiley-Interscience Publication

**JOHN WILEY & SONS, INC.**

This book is printed on acid-free paper. ∞

Copyright © 2002 by John Wiley & Sons, Inc., New York. All rights reserved.

Published simultaneously in Canada.

For ordering and customer service, call 1-800-CALL-WILEY.

*Library of Congress Cataloging-in-Publication Data Is Available*

ISBN 0-471-22290-9

Printed in the United States of America

10 9 8 7 6 5 4 3 2 1

# CONTENTS

# PREFACE

This book provides a new way to understand the subject of electronics. The central theme is that all electrical phenomena can be explained in terms of electric and magnetic fields. Beginning students place their faith in their early instruction. They assume that the way they have been educated is the best way. Any departure from this format just adds complications. This book is a departure—hopefully, one that helps.

There are many engineers and scientists struggling to function in the real world. Their education did not prepare them for handling most of the practical problems they encounter. The practitioner in trouble with grounds, noise, and interference feels that something is missing in his education. The new engineer has a very difficult time ordering, specifying, or using hardware correctly. Facilities and power distribution are a mystery. Surprisingly, all these areas are accessible once the correct viewpoint is taken. This book has been written to provide a better introduction to the field of electronics so that the parts that are often omitted can be put into perspective.

The book uses very little mathematics. It helps to have some background in electronics, but it is not necessary. The beginning student may need some help from an instructor to fill in some of the blanks. The practicing engineer will be able to read this book with ease.

Field phenomena are often felt to be the domain of the physicist. In a sense this is correct. Unfortunately, without a field-based understanding, many electronic processes must remain mysteries. It is not necessary to solve difficult problems to have an appreciation of how things work. It is only necessary to appreciate the fundamentals and understand the true nature of the world.

To illustrate the problem, consider an electric field that is constant everywhere. Place a conducting loop of wire at some crazy angle in this field and ask a question: What is the shape of the new field? This is a very difficult problem even with a great deal of computing power. Now, have the field change sinusoidally and consider current flow and skin effect and the problem really gets difficult. The ideas are important, but the exact answer is not worth worrying about. Connecting wires and components to form circuits is standard practice. These conductors modify the fields around them. This is the same nasty problem, and again it does not need an exact solution. What is needed is an understanding of what actually takes place. Circuit theory does not consider this type of problem.

Most students in electronics spend a great deal of time with circuit theory. The viewpoint of circuit theory is to treat lumped-parameter models. Circuit theory provides an excellent way to predict the behavior of a group of components. The mathematics is very straightforward. Field theory, on the other hand, provides very little in terms of simple answers. Most practical problems cannot be approached by field theory, and yet circuit theory and field theory are inseparable. Circuit theory has no way to handle component size or orientation. Circuit theory, with its zero-ohm connections, avoids any reference to loop area, common-impedance coupling, or common-mode coupling. It fails to reference radiated energy from any source. Circuit theory has its successes, but it also has its failures. Field theory has its place, too, and yet it fails, as there is no convenient methodology.

Educators are oriented toward problem-solving courses. Circuit theory fits this model, as it lends itself to solving many practical problems. Electricity and magnetism courses are more difficult, and only very simple geometries can be approached. The mathematics of vector fields, complex variable, and partial differential equations are not for the faint of heart. This leaves the practicing engineer with one solution. Drop physics and concentrate on circuit theory to provide answers. The circuit diagram of a building or the grounding diagram of a power grid is of no help in analyzing interference. These diagrams can be attempted, but they fail to provide a useful approach. They do not fit the textbook models, as they are not lumped-parameter circuits. The engineer is at a loss.

This book allows the student to solve problems by means of simple ratios. In each area, typical practical problems are solved in the text. The student is expected to use this information to work the problem sets. The answers are all worked out in Appendix I. This makes it possible for the engineer or technician out of school to use the book for self-study. It also makes it possible to use the text in school, where problems can be assigned. The teacher can modify the parameters in the problems so that the student must work out the details rather than copy the answers.

This book is not intended to teach circuit theory. It is not a substitute for teaching physics. It is a tool that can be used to connect the two subjects. There is a need to establish an elementary understanding in both areas so that the reader can understand the things that occur in the real world. This is done in the early chapters. The problems that are discussed throughout the book occur frequently. Exact solutions are not attempted. The simplifications that are applied are brought out in the text. These simple approaches provide insight into what can be done to handle practical situations. If students want to study physics or expand their knowledge of circuit theory, many texts and courses are available. This book takes the liberty of choosing important features from both areas in order to provide students with a different view of the electrical world—a view from the bridge between electrical behavior and physics.

*Redwood City, CA*                                                RALPH MORRISON
*February 14, 2001*

# 1 The Electric Field

## 1.1 INTRODUCTION

This book is written to bring together two topics: *circuit theory* and *field theory*. Electromagnetic field theory is an important part of basic physics. In school it is usually taught as a separate course. Because physics is a very mathematical subject, the connection to everyday problems is not emphasized. Circuit theory, by its very nature, is very practical. It provides a methodology that connects with the many problems that students will encounter in practice. It is natural for most technical people to reinforce circuit concepts and push basic physics into the background.

Circuit theory is not a match for describing the nature of a facility, the interconnection of many pieces of hardware, or the power grid that interfaces each piece of hardware. In circuit theory the emphasis is on components, not on such items as facilities or power distribution. A building or a power grid is not a topic for discussion in a physics course, as these areas are far too complex to consider. Basic physics can handle only very simple geometries, not buildings. Given an interference problem, an engineer defaults to circuit theory and circuit diagrams, as this is where he or she is usually most successful. The circuits that might be considered for a facility usually do not communicate well and bring little understanding to the problem. The fact that there are no actual circuit components to consider is just one of the problems.

Circuit theory is a very powerful tool. If the right circuits are considered, the answers can be meaningful. In this book we place the concepts of fields into every aspect of circuit behavior. Every component functions because of internal or external fields. A facility has its own fields, and these fields enter into every circuit. When all the fields are considered, many problem areas become clearer. A solution may require changing the geometry of a system to limit the influence of the extraneous fields. Circuit theory is still used, but the influence of the environment becomes a part of the design. In effect, field theory brings geometry into circuit design. Experienced designers understand how important geometry can be to circuit performance.

Fields are fundamental even in static circuits, and this is where the first chapter starts. All circuits function through the motion of field energy, and this idea must be considered at all circuit speeds. This includes batteries, utility power, audio, radio frequencies, and microwaves. Fields are needed to operate every circuit component, and conductors are needed to bring fields to each component. This means that the flow of field energy, to every component describes performance. The environment also includes field energy and this energy cannot be ignored. Understanding this fact makes it possible to design practical products.

Today's circuits operate at very high speeds. The demand to process vast amounts of data in very short periods is ever present. To understand high-speed problems, it is necessary to start slowly. The fields involved in all electrical phenomena are the same. In the first chapter we treat static charge and the concept of voltage. These very elementary ideas lay the foundation for understanding circuit behavior at all speeds. In later chapters, when the fields are changing more rapidly, the problems of radiation are discussed. All circuits, including the lowly flashlight, are explained using the same physics. This is where the book starts: fields, batteries, and resistors.

## 1.2  CHARGE

In very dry weather, rubbing a comb through one's hair will cause *static electricity*. The rubbing has removed some of the electrons from the surface of the comb. This group of electrons is said to be a *charge*. Since electrons are negative charges, the comb is left positively charged. Thus the absence of electrons is also considered a charge. In a clothes drier, where clothes are rubbed together and against the walls of the tumbler, charges are moved from one surface to another. This condition can reach a point where the electrical pressures in the dryer space remove outer electrons from air molecules. This process is known as *ionization*. The motion of electrons between molecules causes a glow that can be seen in dim light. The same thing happens in the atmosphere when falling raindrops strip outer electrons from air molecules. Raindrops carry these electrons to earth, leaving a net positive charge. This ionization in the air builds in intensity until there is breakdown or lightning. The electrons now have a path to return to the clouds where they originated.

Normally, the surface charges on an object are balanced by opposite charges located inside the atoms (protons). This means that on average, physical objects are neutrally charged. When electrons are moved from one body to another, the object receiving electrons is charged negatively and the object giving up the charge is said to be charged positively. A steady charged condition is not normally found in nature. In time, any accumulation of charges will dissipate and a neutral condition will return.

The idea of having a positive charge as a counterpart to the negative charge (a group of electrons) is appealing. In the real world a positive charge is usually the absence of electrons. It really makes no difference if we use the concept of positive charges as opposed to the absence of negative charges. In a semiconductor, electrons move inside a crystal lattice. When they move, they leave a hole (a vacant space). In effect, negative charges move one direction and holes move in the opposite direction. The holes behave very much like positive charges. An example in real life might be people seated in an auditorium. Assume that a row has an empty seat at the right end. If the people move one at a time to sit in the empty seat, the people move right but the empty seat moves left.

The number of electrons on the surface of any metal or insulator is extremely large. For most electrical activity the percentage of electrons that are moved is infinitesimal, yet the effects can easily be observed and measured. The letter $Q$ is used to represent positive charge—usually a depletion of electrons. The unit of charge is the coulomb (C). In some cases this unit is extremely large. A more practical unit is the microcoulomb ($\mu$C) one millionth of a coulomb. One electron has a charge of $-1.6 \times 10^{-19}$ C.*

## 1.3  ELECTRICAL FORCES ON CHARGED BODIES

It is relatively easy to perform tests on charged objects. Procedures exist that can remove or add charges to objects. Rubbing a hard-rubber wand with a silk cloth is one technique. Touching this charged rod to small insulators that hang on a string can transfer charge onto the balls. Two balls will repel each other if they both have positive or negative charges. When the charges are of opposite sign, the balls will attract. These forces are between charges, not "between" the matter in the ball. The larger the charge that is added, the greater the force. If the insulating balls are replaced by very small, lightweight metallized spheres hanging on insulating threads, the results are the same (Figure 1.1). The amazing thing here is that there is a force acting at a distance. The forces exist whether the spheres are in air or in a vacuum. On a perfect insulator, the forces cannot move the charges around on the object. A nearly ideal insulator would be glass. On a metal sphere the excess charges spread out over the surface as like charges repel each other. This is the same force that repelled the two spheres in Figure 1.1. These charges cannot leave the sphere, as there

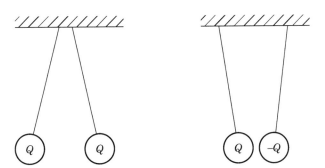

**FIGURE 1.1**  Charged metallized spheres.

---

*The abbreviations used in this book are listed in Appendix III at the back of this book.

is no available conductive path. This force between charges is one of the fundamental forces in nature. It is one of the forces that hold all molecules together in all matter.* Gravity is another fundamental force that acts at a distance. It is a weak force because it takes the mass of the entire earth to attract a person with a force equal to his or her weight.

## 1.4  ELECTRIC FIELD

When forces exist at a distance, it is common practice to say that a force field exists in space. In this case, the force field is called an *electric field* or *E field*. This field is represented by field lines drawn between charged objects. These symbolic lines connect units of positive charge (the absence of electrons) with units of negative charge. When more charges are involved, convention says that there are more lines (Figure 1.2).

The nature of the field is determined by placing a small test charge in the field. Note that the test charge must be small enough not to change the nature of the field that it is measuring. (This test charge is truly hypothetical. It may not be realizable except as a thought experiment. It does take a bit of faith to accept this idea.) The test charge experiences forces that have both magnitude and direction. The lines are drawn so that a small arrow on the line points in the direction of the force. After the lines are drawn, it can be determined that the forces are greatest near the charged objects where the lines get close together. The *E* field exists through all space, not just on the lines. Thus these lines are

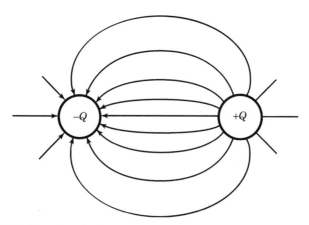

**FIGURE 1.2**   Electric field lines between oppositely charged spheres.

*There are forces between the outer electrons and the protons in the atom's nucleus. When molecules are formed there are binding forces between atoms that are controlled by the electrons in the outer shells of the atoms. These same forces help to bind molecules together in solids and liquids.

only a representation. At every point in space, the force has a magnitude and a direction. This is properly known as a *vector field.*

The spheres in Figure 1.2 are relatively close together. If one of the spheres is moved very far away, the $E$ field on the remaining sphere still exists. The lines leaving the near sphere will be evenly spaced around the sphere. This means that the charges are spaced uniformly on the sphere's surface. This uniform spacing is unique to a sphere. On any other conductor shape the charges will arrange themselves so that the resulting field stores the least amount of energy. The idea of field energy storage is discussed in more detail in later sections.

## 1.5  WORK

In physics, the definition of work is force times distance, $f \times d$, where $f$ and $d$ are in the same direction. A good example of mechanical work involves lifting a bottle of water into a storage tank. If the tank is 25 feet (ft) high and the water weighs 1 pound (lb), then the work expended per bottle of water is 25 ft-lb. In the intervening space the work is 1 ft-lb for every foot in elevation. In the case of a test charge in an electric field, work is done in moving this charge between the two charged bodies. A force is required to move the test charge along any field line. The force for short distances along this line is nearly constant. The work over any short interval on this line is the $E$ field intensity times this short distance. The total work along the entire path is the sum of all the bits of work. The work done in moving the bottle of water is stored as potential energy. When the water is released, it can do work as it falls: for example, it could turn a turbine. The same thing happens when charges are moved in a field and added to a conductor. The work that is done on the charge is stored and is available to do work when it is released. It will turn out that this work is actually stored in the electric field. Work in this case is the process of adding to or subtracting from the electric field. Once the energy is stored, it can be used at a later time. This use of stored energy is an important topic in the book.

In the case of the bottle of water, the path taken by the bottle does not change the amount of work that must be done. The same thing is true of the unit charge. No matter what path is taken, the work required to move the unit charge between the charged bodies is the same.[†] This type of field is said to be *conservative.* Gravity is also a field phenomenon. The gravitational field and the electric field are both examples of a conservative field. Later we discuss the magnetic field, which is not conservative.

---

[*]The $E$ field is often represented by a line with an arrow. The length of the line represents the intensity and the arrow shows the direction of the force on a test charge. This arrow is only a representation of the intensity and direction at a point in space.
[†]In calculating work, the force and the distance moved must be in the same direction. If other paths are taken, the angle between the force and direction of motion must be a part of the calculation.

Free electrons in a vacuum are accelerated by an electric field. This is analogous to a mass above Earth accelerated by gravity. In a conductor the electrons are also accelerated, but they keep bumping into molecules. This means that on average they do not accelerate. This motion of charge is a current and it takes a continuous $E$ field on the inside of the conductor to keep charges moving at an average velocity. In all the discussions above, the fields are static and it is assumed that the charges are not moving (the exception being the test charge).

## 1.6  VOLTAGE

The fundamental definition of voltage relates to the work required to move a unit of charge between two points. In this case the unit of charge is our test charge. By convention, the unit of charge is positive. The amount of work does not in any way require a reference level. To lift water 25 ft, the amount of work required is the same whether this work is done at sea level or at 5000 ft. (This assumes that the gravitational force is constant.) The same is true in the electric field. The work we are interested in involves moving the test charge between the two bodies. The work required is measured by the potential difference. It is correct to say that the work per unit charge is the voltage difference. The words *voltage* and *potential* are thus used interchangeably.

Any point can be selected to be the zero of potential. If a remote point is selected, work may be required to get the test charge to the first body. If this work is 10 volts (V), then the work required to get to the second body may be 5 more volts. The potential difference between the two bodies is simply 5 V. There is no place that can be called the absolute zero of potential. It is misleading to believe that such a point exists. It will be obvious as we proceed that potential differences are our main concern.

When the force is positive and the test charge is positive, positive work is done in moving this charge. This work is actually stored in the $E$ field as potential energy. When the charge is allowed to return to its starting point, a bit of potential energy is removed from the field.

The abbreviation mV stands for millivolt (0.001 V), $\mu$V stands for microvolt (0.000001 V), and kV stands for kilovolt (1000 V). The range of values that is encountered in practice is large. Writing lots of zeros before or after the decimal point is really an inconvenience. The circuit symbol for a source of voltage is a circle with the letter $V$ in the center.

## 1.7  CHARGES ON SURFACES

An $E$ field exerts forces on charges. If these charges are on a conductive surface, they will try to move apart. The small metallized spheres we used in Figure 1.2 held charges, which generated an $E$ field. These charges were distributed over the conducting surface. Since the charges were at balance and not moving, we conclude that there cannot be a component of the $E$ field (a force) directed along the surface of the sphere. If there were a tangential $E$ field, there would be current flow. This is impossible because we have postulated a static situation. This means that any $E$ field that touches the conductive surface must have a direction that is perpendicular to the surface. These $E$-field lines must terminate or originate on surface charges. This $E$ field cannot move these charges, as the electrons cannot jump off the surface into the surrounding space. Also note that an $E$ field cannot exist inside the metal, or charges would be moving to the surface. Again remember that this is a conductive material, and an $E$ field would imply a current moving to the surface from within. These arguments lead us to three important conclusions:

1. For there to be a voltage difference, charges must be present. These charges result in an $E$ field.
2. In electric circuits with potential differences, charges exist on the surfaces of all conductors. In a static situation, the $E$ field touching a conductive surface has a direction perpendicular to the surface. The field does not extend into the surface. In Figure 1.2 the field lines terminate on charges at the surface of the spheres. Note that most of the lines terminate on the facing sides of the spheres. This means that the charges do not spread out evenly.
3. Charge distributions are not necessarily uniform on a conductive surface. In a static situation, the potential along the surface is constant. This means that the work required to bring a test charge to the conductive surface is the same for all points on the surface.

In Figure 1.3 the field pattern for two conductors over a conductive plane is shown. Conductor 1 is at a potential of 1 V and conductor 2 is at a potential

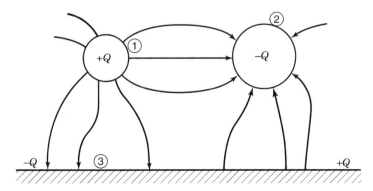

**FIGURE 1.3**   Field pattern of three conductors.

of −2 V. This means it takes 1 V of work to move a unit charge from the conductive plane to the surface of the first conductor. By convention the field lines have arrows showing that they start on positive charges and terminate on negative charges. Consider the conductive plane as the reference conductor and consider it to be at 0 V. It takes 2 V of work to move the unit charge from conductor 2 back to the conductive plane. A conducting plane is sometimes called a *ground plane* or a *reference plane*. It is important to note that the ground plane has areas with positive and negative charge accumulations on its surface. Also remember that no work is required to move a unit charge along this reference surface. The entire surface is at one potential, which is defined as zero. In this example the two conductors could be round wires used to connect points in a circuit. The voltages on the conductors might represent signals at one point in time. The reference conductor could be a metal chassis or a metallized surface on a printed circuit board.

## 1.8 EQUIPOTENTIAL SURFACES

In Figures 1.2 and 1.3 the geometry is simple and it is easy to draw the *E*-field lines. In most circuits, the conductor geometries are far too complex to consider drawing field patterns. This does not stop nature, as the fields do exist. When there are voltages, there are charge distribution patterns and there are fields. This fundamental idea is often forgotten. So far we have discussed the *E* or force field. The next step is to discuss the associated equipotential surfaces.

When a unit test charge is moved from one surface to another, the work required is the potential difference. As the test charge is moved, it is possible to note points of constant work (constant potential). A plot of all points that are at the same potential is an equipotential surface. This is equivalent to climbing a mountain and noting points of equal elevation. In Figure 1.4, two spheres are shown with intermediate equipotential surfaces. Of course, the conducting spheres themselves are equipotential surfaces. In the space between the spheres, these equipotential surfaces are everywhere perpendicular to field lines. Moving a test charge along these new surfaces requires no work. The figure shows that the equipotential surfaces are close together near the spheres. This is the same thing as saying that the mountain is getting steeper as we near the summit. The work required to move the test charge a unit of distance is greatest near the surfaces. This is where the field lines are closest together. This is where the field is said to have its highest gradient.

## 1.9 FIELD UNITS

In the previous figures the *E*-field lines are curved and not equally spaced. This implies that the intensity of the *E* field changes over all space. As noted

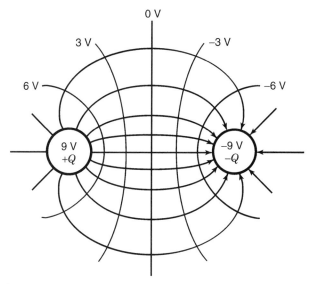

**FIGURE 1.4**   Equipotential surfaces perpendicular to field lines.

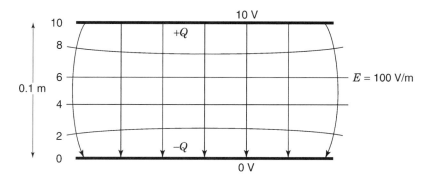

**FIGURE 1.5**   $E$ field between parallel conductive plates.

earlier, where the $E$ field lines get closer together, the force on a test charge increases. A simpler field pattern results when charges are placed on two parallel conductive planes as in Figure 1.5. The $E$ field in the central area are straight lines. This means that the force on a test charge is constant at any point between the two surfaces. If the distance is 0.1 meter (m) and the potential difference is 10 V, the $E$ field times 0.1 m must equal 10 V. In other words, the $E$ field must be expressed as 100 V/m. In equation form, 100 V/m × 0.1 m = 10 V. Thus the $E$ field has units of volts per meter. Two parallel conducting surfaces form what is known as a *capacitor*. More will be said about capacitors in later sections.

## 1.10   BATTERIES—A VOLTAGE SOURCE

Energy can be stored chemically. When there is a chemical reaction, energy is released. In an explosion, this energy can be released as heat, light, and mechanical motion. In some arrangements, chemical energy can be released electrically. A battery is an arrangement of chemicals that react when the active components are allowed to circulate their electrons in an external circuit. The energy that is stored chemically is potential energy that is available to do electrical work. In rechargeable batteries the chemistry is reversible and energy can be put back into the battery.

The terminals of the battery present a voltage to the world. This is electrical pressure trying to move electrons so that the chemicals in the battery can attain a lower energy state. This is analogous to water pressure in a water tank where the water is trying to get to a lower energy state. This water pressure is no different from the voltage between two oppositely charged conductors in space. There is an $E$ field between the terminals of the battery. If this is a 12-V battery, it takes 12 V of work to move a unit of charge between the two terminals. This work is independent of the path taken by the test charge. This includes a path through the heart of the battery. The $E$ field cannot be seen, but it is there. This field extends right into the battery, where the atoms are under pressure to release their external electrons.

In Figure 1.2 the static charges can be removed and the $E$ field disappears. In the case of the battery, the $E$ field and the associated charges on the conductors will persist until the battery is dead. When charge is allowed to flow through a circuit connected to the terminals, the battery replaces this charge and maintains the electrical pressure. A battery is thus a voltage source that does not sag. It is like being connected to the city water supply. No matter how much water you draw, the water pressure is the same. The $E$ field around battery terminals is shown in Figure 1.6. The positive terminal is called the *anode* and the negative terminal the *cathode*. The charges that are allowed to flow from a battery to a circuit release stored chemical energy. The voltage and associated charge flow constitutes the electrical energy that is flowing from the battery. A connected circuit can convert this energy to heat, light, or sound. In some cases it is radiated. An example of radiation might be a cell phone transmission. In most common circuit applications the energy is converted to heat. Of course, it is possible to store some of this energy in $E$ fields within a circuit. More will be said about this later.

Batteries are usually formed from basic cells. A typical flashlight battery is such a single cell. The single-cell voltage in most size A and D batteries is 1.1 to 1.5 V. Different battery materials develop different voltages. To obtain higher battery voltages, basic cells are placed in series. The cell connections are made internally, and the connections are not available for external connections. A 12-V automobile lead acid battery is constructed with six such cells in series. Each internal plate forms a cell that develops a voltage of about 2.0 V. Batteries can be connected in series to increase the available voltage. This

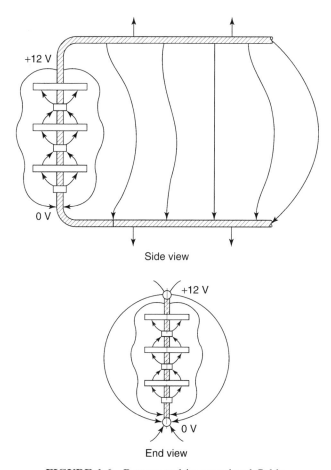

**FIGURE 1.6**   Battery and its associated fields.

series arrangement will work even if the batteries have different voltages. Batteries cannot be paralleled unless the batteries themselves are identical. This
parallel arrangement can be used to provide additional current capacity. When
considerable power is involved, very careful monitoring of the batteries is
necessary.

## 1.11   CURRENT

The motion of charge is current. The unit of current is the ampere. The letter
symbol for current is $A$ or sometimes $I$. A source of current is often represented
by the letter $I$ in a circle. The smallest charge is an electron. In most practical
circuits the number of electrons that constitute current flow is so large that
it makes little sense to consider the individual electrons. There are cases,

however, where individual electrons are counted, such as in a photomultiplier. For our discussion, current flow is continuous and the effect of individual electrons is not considered. A steady current is a continuous stream of charges that flow past an area. A coulomb of charge passing by in 1 second is defined as 1 ampere (A). In other words, a coulomb per second is an ampere. In equation form, $Q/t = A$. In the power industry, an ampere is a small unit. In an electronics circuit it is a big unit. For this reason, smaller units of current are a convenience. The abbreviation mA stands for milliampere (0.001 A) and the abbreviation $\mu$A stands for microampere (0.000001 A).

The positive terminal of a battery is a source of current flowing out of the battery. This direction is a convention only. The actual flow of negative charges is in the opposite direction. This may seem confusing at first, but it is how the world of electricity developed. Historically, there was an assumption that moving charges are positive and the convention has persisted. Electrons are attracted to the positive battery terminal, but by convention, current flows out of this terminal.

Current does not ordinarily flow in air. It can flow easily in conductors such as copper or iron. Plastics and glass are examples of very poor conductors. Conductors for electrical wiring are available in many configurations, all the way from power lines to circuit traces on a printed circuit board. When conductors are attached to a battery, the field across the terminals is extended out on the conductors. This means that charges now exist on the surface of these added conductors. These charges move out on the conductors looking for a path that will allow them to work their way from the anode to the cathode (current is seeking a path to flow from the cathode to the anode).

A direct conducting path between the cathode and anode will destroy the battery. This direct path simply shorts out the battery. The circuits that are normally connected to the terminals will drain charge at a rate that the battery can supply for a useful period. A car battery might be able to supply 1 A for 60 hours, for example. A flashlight battery might be able to supply 100 mA for 10 hours.

## 1.12   RESISTORS

A resistor is a controlled limited conductor. When electrical pressure is applied across its terminals, a limited current will flow. The water pipe analogy can serve to illustrate the point. Consider a water hose connected to a cylinder full of packed sand. The amount of water that could flow through the cylinder will depend on the length of the cylinder, the cross-sectional area, the size of the grains of sand, and the water pressure. This cylinder is in effect a water flow restrictor. The electrical form of this restrictor is a resistor. One type of resistor is made from a mixture of powdered carbon and a nonconductive plastic filler. This mixture in compressed form constitutes a resistor. The resistance can be controlled by varying the ratio of filler to carbon. This controlled mixture

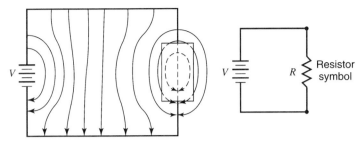

**FIGURE 1.7** *E* field along a resistor.

is fused together under pressure to form the resistive element. This element is encased in an inert housing that is marked with a color code. Conductive leads that contact the resistive element at both ends are molded into the body of the resistor. This briefly describes a carbon resistor, which is just one of many resistor types commercially available. The letter symbol for a resistor is a capital *R*. The electrical symbol for a resistor is shown in Figure 1.7. When an electric field is impressed across a resistor, a limited current flows. For a practical resistor, if the electrical pressure is doubled, the current flow is doubled. Figure 1.7 shows how the *E* field distributes itself along and through the resistor. The field, in effect, pushes charges along over the entire resistive path. Note that there is an *E* field inside the resistor. Also note that some of the field bypasses the resistor. This field will be important in later discussions.

Resistance to current flow has units of ohms. The symbol for ohm is the capital Greek letter omega ($\Omega$). A resistance of 1 ohm will limit the current flow to 1 ampere when the electrical pressure is 1 volt. Typical circuit resistors are usually much greater than 1 $\Omega$. A 1000-$\Omega$ resistor will limit the current flow to 1 mA for an electrical pressure of 1 V.

The linear relationship between resistance and current flow is known as *Ohm's law*. Double the voltage and the current flow doubles. Double the resistance and the current flow halves. In equation form, the relationship is given as $I = V/R$, where *I* is in amperes, *V* is in volts, and *R* is in ohms.

## 1.13  RESISTORS IN SERIES OR PARALLEL

When resistors are placed in series, the total resistance simply adds. A 1000-$\Omega$ resistor in series with 2000 $\Omega$ is simply 3000 $\Omega$. When resistors are placed in parallel across a voltage source, the total current that flows is the sum of the two individual currents. Using this new current and the applied voltage, an equivalent resistance can be calculated. For example, a voltage of 10 V across a 1000-$\Omega$ resistor results in a current of 10 mA. Ten volts across a 2000-$\Omega$ resistor results in a current of 5 mA. For these resistors in parallel, the total current is thus 15 mA. The equivalent resistance value is thus the voltage divided by the current (expressed in amperes). The answer is simply 10 divided by 0.015, or 666.6 Ω.

Another way to look at resistors is to consider how well they conduct. Resistors resist current flow and thus their name. It also makes sense to measure a resistor's ability to conduct current in terms of conductance. It is easy to see that the higher the resistance, the lower the conductance. This simply means that resistance is the reciprocal of conductance. The unit of conductance is the siemens (S). A resistance of 2 Ω is said to have a conductance of 0.5 S. A 1000-Ω resistor has a conductance of 0.001 S, which can also be written as 1 mS.

Using the concept of conductance, when resistors are placed in parallel their conductances add. In the preceding example, the 1000-Ω resistor has a conductance of 1 mS. The 2000-Ω resistor has a conductance of 0.5 mS. The combined conductance is simply the sum of the two conductances, or 1.5 mS. This can be converted to ohms simply by taking the reciprocal of the conductance in units of siemens. Our 1.5 mS can be written as 0.015 S. The reciprocal of 0.015 S is 666.6 Ω, the same answer as before.

When two equal resistors are placed in series, the voltage across each resistor is one-half the total voltage. This means that the $E$ field is reconfigured around the resistors so that the work required to move a unit charge around or through each resistor is one half the total voltage. If the resistors are placed at right angles to each other, the $E$ field takes on a new shape to make this happen. A typical field pattern is shown in Figure 1.8.

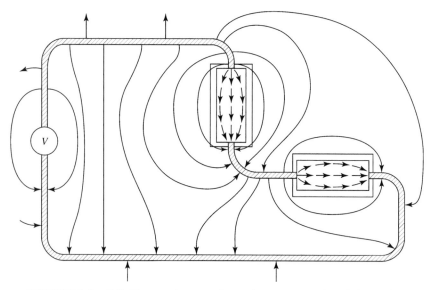

**FIGURE 1.8** $E$ field around two series resistors mounted at right angles.

## 1.14  *E* FIELD AND CURRENT FLOW

The *E* field in Figure 1.7 enters the heart of the resistor to push the charges through. The battery and the resistor form a very simple circuit. When there is current flow, even in a conductor, some *E* field is needed to push the charges along. In effect, the conductors that connect the resistor are also resistors, but their resistance is very low. Typical connecting conductors might have resistances below 0.001 Ω. In Figure 1.7 the *E* field appears perpendicular to the conductors. To push the charges along, a small component of the *E* field exists inside the connecting wires in the direction of current flow. This means that the *E* field is not exactly perpendicular to the conductor and it leans forward just a little. There are still charges on the outside surface of the conductors, as most of the *E* field is perpendicular to the surface.

The *E* field component that is inside the conductor can be calculated as follows. If the current is 10 mA and the resistance is 1 mΩ, Ohm's law requires a voltage drop of 0.000010 V. If the path length is 0.1 m, the *E* field is 0.00010 V/m. Compare this with the *E* field between the conductors if the conductor spacing is 0.1 m. The *E* field is approximately 10 V per 0.1 m, or 100 V/m. In other words, the component of the *E* field in the conductor is one millionth of the *E* field between the conductors. Of course, this would change if a larger current were to flow.

Charges do not move in a practical conductor unless there is an *E* field to push them along. In a vacuum, electrons are unimpeded and accelerate across the available space. They give up their energy on impact at the end of their journey. This is what happens in a vacuum tube when the electrons strike the plate. Without a vacuum, the moving charges constantly collide with atoms. On average, they attain a fixed velocity. This steady flow of charge is called a *dc current*. The abbreviation *dc* stands for direct current. A steady or dc current that flows in a conductor flows uniformly though the entire conductor.* It does not flow on the surface. In a resistor the charges flow uniformly in the resistive element.

## 1.15  PROBLEMS

1. A 10-V battery is connected to two conductors. What is the *E*-field intensity when the conductors are separated by the following distances: 1 m, 10 cm, 1 cm, and 1 mm?

2. A 10-kΩ resistor element is 1 cm long. The *E* field is 20 V/m. What current flows in the resistor?

3. A 2000-Ω resistor is placed in series with a 3000-Ω resistor. What is the total resistance? What is the parallel resistance? What is the parallel conductance expressed in siemens?

---

*The charges that move on the inside of the conductors at dc are distinct from the static charges that reside on the surface as the result of external *E* fields.

**4.** A conductor has a resistance of 0.002 $\Omega$/cm. How much current flows if the $E$ field parallel to the conductor is 0.0001 V/m?

**5.** Two conductors carrying signals over a ground plane are at 2 V and $-3$ V. How much work is required to move a test charge between the two conductors?

**6.** Two resistors with conductances of 1 mS and 3 mS are placed in series. What is the total resistance? What is the resistance if they are placed in parallel?

**7.** Four resistors of 1000 $\Omega$ each are arranged in a square. What is the resistance across either diagonal? If 10 V is placed across one diagonal, what is the voltage across the other diagonal?

**8.** In problem 7, assume that one of the resistors increases by 1%. If 10 V is placed across one diagonal, what is the voltage across the other diagonal?

**9.** A 12-V battery sags 0.1 V when supplying 10 A. What is the internal resistance of the battery? (*Hint*: The internal resistance is a series resistor.)

**10.** The battery in problem 9 is being charged at 2 A. Assume that the internal battery voltage stays at 12 V. What voltage must be supplied by the charger?

## 1.16  ENERGY TRANSFER

The charges that move in a resistor through an $E$ field convert their potential energy to heat. This happens because the molecules that are hit by these moving charges take on a higher average velocity. This increased motion of molecules is simply heat. The energy is first stored chemically in the battery. Some of this energy is moved into the $E$ field. Charges moving in this field take energy from the field and convert it to energy of motion. This motion is transferred to the molecules of the resistor as heat. The energy that is removed from the $E$ field is resupplied by the battery. The battery heats the resistor, but there are obviously several intervening steps involved.

This transfer of energy via the $E$ field is not the entire story. How energy moves along conductors is explained in detail in Section 3.10. The important idea here is that the energy heating the resistor involves the $E$ field, which was generated by the charges emanating from the battery.

The work to move a unit charge between two points in a field is simply the voltage $V$. When many units of charge are involved, the total charge is simply $Q$. When a charge $Q$ moves through a potential difference, the total work involved is the product of charge $\times$ voltage, $QV$. Assume that this work is stored as energy. This energy is the product of $QV$ and has units of joules (J). The rate at which work is done is a measure of power. This is simply charge $\times$ voltage $\div$ time. When 1.0 J of work is done in 1 second (s), the power is said to be 1 watt (W). In equation form, power equals $QV/t$. From Section 1.11, current is simply charge per unit time, or $Q/t$. By letting $Q/t = A$, the power becomes volts × amperes, or $P = VA$. If the voltage is 10 V and the resistor is 1000 $\Omega$, the current is 10 mA. The product of voltage × current is $10 \times 0.01$, or 0.1 W. This means that 0.1 J of energy is dissipated in the resistor in each second.

## 1.17   RESISTOR DISSIPATION

When current flows in a resistor, it dissipates heat. The temperature rise will depend on its radiating surface and on how the resistor is cooled. In most commercial applications it is a good idea to limit dissipation to about one-half the rating. In practice, most resistors are used far below their ratings. It is convenient to use resistors of one size and ignore the issue of ratings except where it is obviously a problem. In an integrated circuit, the resistors are individually designed.

The power dissipated in a resistor is simply $IV$, where $I$ is in amperes and $V$ is in volts. For example, the current that flows in a 100-$\Omega$ resistor with 10 V across its terminals is 0.1 A. The dissipation in watts is 10 V × 0.1 A, or 1 W. Three parameters are involved: resistance, current, and resistance. When two parameters are known, the power dissipated can be calculated. Power is $I^2R$ or $V^2/R$ or $VI$.

## 1.18   PROBLEMS

1. Ten volts is impressed across a 100- and 200-$\Omega$ resistor in parallel. What is the power dissipated in each resistor? What is the total power dissipated?

2. Ten volts is impressed across a 100- and a 200-$\Omega$ resistor in series. What is the power dissipated in each resistor? What is the total power dissipated?

3. A charge of 4 C flows across a potential of 4 V in 8 s. What is the energy dissipated in joules? What was the power level during this interval of time?

4. Four $\frac{1}{4}$-W 100-$\Omega$ resistors are available to dissipate energy. How many configurations will dissipate equal power in each resistor? What configuration will accept the greatest voltage? What configuration will accept the smallest voltage?

5. In problem 4, assume that the maximum dissipation per resistor is $\frac{1}{8}$ W. What are the voltages applied to each configuration? How many resistance values are available?

6. A switch connects a 10-V battery for 2 s and disconnects it for 3 s. If this switching cycle is repeated every 5 s, what is the average power dissipated in a 40-$\Omega$ resistor? What wattage rating would you select for the resistor?

**7.** A 24-V battery is reversed in polarity every 2 s. What is the power dissipated in a 48-$\Omega$ resistor?

**8.** How many joules does a 12-V car battery store if its rating is 60 amperes-hours?

**9.** In problem 8, a resistor of what value will discharge the battery in 24 hours? Assume a steady voltage. What wattage will be dissipated in the resistor?

**10.** A headlamp on a car dissipates 200 W. What is the current flow in the headlamp? Assume a 12-V battery. What is the current if the battery is 6 V?

## 1.19   ELECTRIC FIELD ENERGY

When charges are distributed onto conductors, electric fields exist. Consider Figure 1.5, where there are two parallel plates relatively close together. If opposite charges are placed on the two plates, an $E$ field will exist between the plates. The field lines start on the top surface and terminate on the bottom surface. When charge is moved from the bottom surface to the top surface, work must be done on the charge. This work is volts per unit charge. If the edges are avoided, the $E$ field has the same intensity at all points between the plates. Thus $V$ is equal to $Ed$, where $d$ is the spacing in meters. When charge is moved across the distance $d$, the work done on the charge does not make any physical change to the plates. The only change that can be observed is that the $E$ field is increased. The work done in moving the new charge must therefore be stored in the increased $E$ field. The $E$ field increases over the total volume between the plates. It is correct to say that an $E$ field stores energy per unit volume of space. In a static situation the $E$ field does not exist inside conductors. This means that the energy can exist only in the space between conductors. No energy is stored in the conductors.

In most conductor configurations the $E$-field intensity varies in space. To calculate the total field energy, the space must be divided into very small volumes where the $E$ field is relatively constant. The total field energy is the sum of the energies in all the small volumes. In general, this calculation is very difficult to make. The important thing here is the concept of field energy stored in space.

The $E$ field is proportional to the charge on the plates. If the plate area is one square meter, a coulomb of charge produces a field of 1 V/m. To move charge between the plates requires work. To calculate the total work, the charges must be moved in small increments. The first elements of charge require very little work. As the field increases, the work per unit charge also increases. It turns out that the total work required to move a net charge between the two plates is $\frac{1}{2}EV$. But $V$ is equal to $Ed$. Therefore, the energy in the field is $\frac{1}{2}E^2d$. Remember, this is the energy stored for plates 1 m on a side. For an area

$A$, the energy stored is proportional to area or $\frac{1}{2}\varepsilon_0^2 Ad$. Since $Ad$ is volume, the energy in joules is $\frac{1}{2}\varepsilon_0 E^2 V$, where in this term $V$ stands for volume. The proportionality factor $\varepsilon_0$ is needed to make the units come out correctly. The value of $\varepsilon_0$ is discussed later.

The important thing to understand is when there is an $E$ field, there is energy storage. This fact is often ignored and can be a source of trouble. In Section 1.16 the idea of power was introduced. Energy flowing over some time period represents power. Energy cannot be moved in zero time, as this takes infinite power. If 0.01 J is dissipated in 10 milliseconds (ms), the power level is 1 W for that period of time. If this same energy is dissipated in 1 ms, the power level during this shorter time is 10 W. It is important to realize that $E$ field energy does not simply disappear. It takes time for this energy to be moved or dissipated. This field energy can be used in several ways. For example, the energy can be transferred to another circuit for storage, it can be used to heat other components, it can be used to create sound, or it can be radiated out of the area. Often, all of these processes are involved.

## 1.20   GROUND AND GROUND PLANES

In Figure 1.3 the concept of a ground plane was introduced. Both positive and negative charges were distributed along this conducting surface. The words ground and *ground plane* are a part of the language of electronics and electrical engineering and they are often a source of trouble. The word *ground* is often associated with an adjective that describes function or use. Expressions such as computer ground, analog ground, clean ground, signal ground, and isolated ground are frequently encountered. At this point these adjectives only confuse the issue, and they will be avoided. The definition we will use is that a *ground plane* is a conductive surface referenced to zero volts.

The earth is an electrical conductor. Connections to earth are required for lighting protection and electrical safety. The earth is considered a form of ground plane, although it obviously is not flat. A ground plane may or may not be connected to earth. An example might be in an aircraft. The framework might be considered a ground plane, but it definitely is not connected to earth. A facility built on a lava bed is insulated from earth because lava is an insulator. The building steel might still be considered a ground.

Electronic systems may have many reference conductors. Each reference conductor serves a specific purpose. Some of these reference conductors may be earthed (connected to earth), and others may be defined by the circuit itself. In any case the concept is to have a conductive surface in the circuit that is considered to be at a zero reference potential. The idealized ground plane we have used so far is a source of positive or negative charge. This surface can supply any amount of charge without difficulty and it remains at the zero of potential. For the ideal ground plane, no work is required to move charges along its surface. It is an equipotential surface.

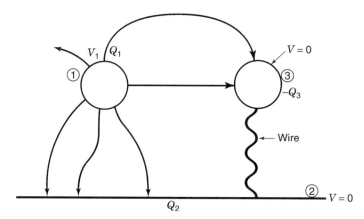

**FIGURE 1.9**   Three-conductor arrangement.

## 1.21   INDUCED CHARGES

The conductors in Figure 1.9 show field lines that terminate on charges. Conductor 3 is grounded (connected to the ground plane) by a small wire. It is at zero potential and it contains a charge of $-Q_3$. If the small connecting wire is cut, the charge on the conductor cannot leave. This charge is said to be induced on the conductor. Work is required to move this conductor in the existing $E$ field. Wherever it is moved, the total charge on its surface remains $Q_3$.

If conductor 3 were initially far removed from conductor 1, it would have no charge on its surface. As it is moved near conductor 1, charges distribute themselves on the surface, but the total charge must remain zero. When the small wire is connected to conductor 3, a charge $-Q_3$ will flow in the wire to this conductor. This is the induced charge shown in Figure 1.9.

## 1.22   FORCES AND ENERGY

The electric field is a force field. These forces are between charges. When charges reside on a mass, the electrical forces are transferred to the mass (see Figure 1.1). Normally, the masses are constrained so that there is no detectable motion. This force field stores potential energy. When a body is moved in an $E$ field, a new field configuration results. This configuration stores a different amount of field energy. The difference is simply the work done in moving the body to new positions. In other words, the forces on a body are related to the change in energy level that results from the motion. The direction of these forces can be determined by noting the direction that causes the greatest change in energy storage. This concept is simple, but the calculation to determine the force and direction on any one conductor is usually very difficult. Forces on conductors in an $E$ field are not generally important in an electronic circuit. The concept of field energy storage is quite important, however. An application of electrostatic force occurs in tweeters used in audio systems. The voltage applied between plates moves the plates, which in turn moves the air. The moving air is the sound we hear. In a cathode ray tube the beam of electrons that writes on the front surface is often deflected by electric fields. The electrons that boil off a heated cathode are accelerated across the viewing tube and give up their energy to phosphors on the surface. These phosphors give up their energy by emitting light that we see.

## 1.23 PROBLEMS

1. The $E$ field in a volume 10 cm $\times$ 10 cm $\times$ 0.1 cm is 1000 V/m. What is the energy stored? (*Hint*: All dimensions must be in meters.)

2. In problem 1, assume that the 10 cm $\times$ 10 cm surfaces are conductors. How much charge resides on these surfaces?

3. Assume that a round conductor runs parallel to and above a ground plane. Sketch the $E$ field lines from the conductor to the ground plane. Assume that the potential of the conductor is 5 V. Draw the equipotential surfaces at 1 and 3 V.

4. The force on a conductor is the change in energy stored divided by the distance moved assuming that the distance is very small. Use this fact to calculate the force between the plates in problem 1. (*Hint*: Calculate the change in energy if the plates are moved 0.01 cm. Use units of meters. The value of $\varepsilon_0$ is $8.854 \times 10^{-12}$. The answer will be in kilograms.)

5. A current flows in a 10-$\Omega$ resistor. What is the average power if the current is 1 A? What is the average power if 10 A flows for 0.1 s every second? What is the average power if 100 A flows for 0.01 s every second?

6. In problem 5, what is the peak power in each case?

## 1.24 REVIEW

The presence of charge implies that there is an electric force field. The work required to move a unit charge in this $E$ field is a measure of potential difference or voltage difference. If there is a potential difference between two conductors, there will be charges on their surfaces. Potential difference is measured in units of volts. There is no absolute zero of potential or zero of voltage. Any conductor can be used as a zero reference conductor. A ground is a conductor that can be used as a reference conductor. A reference conductor may or may not be connected to earth.

One source of potential is a battery where chemical action moves charges in an external circuit. A battery can be used to force a steady flow of charge through a resistor. The charges in the resistor are accelerated in the $E$ field. These charges collide with molecules, which increase their average molecular velocity. This motion is heat. A steady flow of charge is a current measured in amperes. The relationship between resistance, voltage, and current is Ohm's law.

An electric field stores energy. This energy cannot be created or dissipated in zero time. This energy is stored in space, not in conductors. This energy is most dense where the $E$ field is most intense. When energy is taken from an $E$ field, it may be replenished by a voltage source such as a battery. In this case, energy moves from chemical, to $E$ field, to motion of charge, to heat.

# 2 Capacitors, Magnetic Fields, and Transformers

## 2.1 DIELECTRICS

The $E$ field considered in Chapter 1 involved charge on conductors and the fields in the surrounding space. Figure 2.1 shows a dielectric filling the space between two parallel conducting plates. The term *dielectric* refers to the insulating material used in capacitors. Typical materials are Mylar, mica, and polypropylene. When a voltage $V$ is applied between the plates, the work to move a unit charge between the plates is just $V$. The presence of the dielectric increases the charge that is present on the surface of the conductors, and thus the charge per unit voltage is increased. The ratio of charge on the plates with and without the dielectric is known as the *relative dielectric constant*, $\varepsilon_R$. The higher the dielectric constant, the more charge is stored on the plates for a given voltage. The relative dielectric constant for air is 1.0.

In Chapter 1 the $E$-field lines were drawn so that they started on positive charges and terminated on negative charges. When a dielectric is introduced into an existing $E$ field, the field is reduced in the dielectric. This means that there is less field energy stored in this volume of space. The $E$-field discontinuity at the boundary does not imply a charge on the dielectric surface.

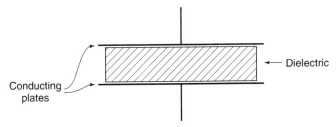

**FIGURE 2.1**   Parallel conducting plates with a dielectric.

## 2.2 DISPLACEMENT FIELD

It is interesting to see what happens when a dielectric occupies part of the space between the two parallel conducting plates. This situation is shown in

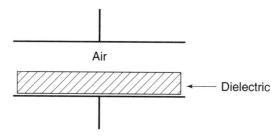

**FIGURE 2.2**   Air and a dielectric between parallel plates.

Figure 2.2. The equipotential surfaces are no longer uniformly spaced across the space. Most of the potential difference appears across the air space. The ratio of the $E$ field in air to the $E$ field in the dielectric is the relative dielectric constant $\varepsilon_R$. It is convenient to consider a new field, one that is not dependent on the dielectric. This field is called the *displacement field* or $D$ field. This $D$ field is generated by charges and it is not a function of the dielectric. To make the units come out correctly, the $D$ field in air is the $E$ field times the dielectric constant of free space. This constant, $\varepsilon_0$, is also known as the *permittivity of free space*. This constant has the value $8.854 \times 10^{-12}$. The $D$ field has units of charge per unit area (coulombs per square meter).

At the interface between air and the dielectric, the $D$ field has the same intensity on both sides of the interface. In Figure 2.2 it is assumed that the dielectric surface is free of charge. If a surface charge did exist, a new $D$ field would start at this surface.

The energy stored in a field is proportional to the $E$ field, not the $D$ field. This is because the $E$ field is still the force field.* A direct measure of the $E$ field inside a dielectric is not practical, so the forces on charges in the dielectric must be inferred. In Figure 2.2 most of the energy is stored in the air space. In Figure 2.1 the energy is all stored in the dielectric, as there is no air space.

When a dielectric is introduced into an existing uniform $E$ field, the field reconfigures itself to store the least possible energy. It is just like water in a puddle. When an object is removed from the puddle, the water level drops and evens out until the water stores the least potential energy. This field pattern when a dielectric is present is shown in Figure 2.3. Liquid dielectrics are often used in high-voltage transformers or in high-voltage power switches. The presence of the dielectric reduces the $E$ field in critical areas and this helps to limit arcing. The liquid can also help in conducting heat out of a big transformer.

---

*The work done in moving charges is stored in the field. The work is equal to the force times distance. The force is proportional to the $E$-field intensity (see Section 1.19).

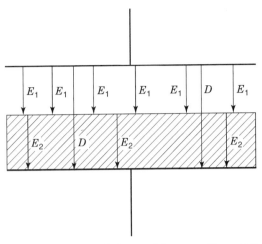

**FIGURE 2.3**   Dielectric inserted into a uniform $E$ field.

## 2.3   CAPACITANCE

The two plates in Figure 2.1 form a capacitor. A capacitor is an electrical component that can store electric field energy. In circuit diagrams the letter symbol $C$ is used to represent a capacitor. (The electrical symbol for a capacitor is identified in Figure 2.5.) A capacitor $C$ is said to have a capacitance. The unit of capacitance is the farad (F). One farad of capacitance means that a charge of 1 coulomb can be stored for 1 volt of potential difference. A typical capacitor has capacitance measured in microfarads ($\mu$F). This is one millionth of a farad. A full farad of capacitance is possible, but it is not the usual circuit component. Capacitors cover the practical range from a few picofarads to thousands of microfarads. A picofarad (pF) is one millionth of a microfarad. The statement $C = 10\ \mu$F means that capacitor $C$ has a capacitance of 10 $\mu$F. It is interesting to note that capacitor values cover a range of over nine orders of magnitude. This is greater than the ratio of an inch to the distance around our earth.

Capacitors come in many shapes and forms, one of which is a foil capacitor. It is made from layers of metalized paper rolled into a tight cylinder. Smaller capacitors are often just parallel layers of metallized dielectrics. Whatever the manufacturing technique, the basic capacitor is a version of Figure 2.1.

## 2.4   CAPACITANCE OF TWO PARALLEL PLATES

*Capacitance* is the ratio of charge stored to the voltage applied. It is a geometric property of conductors and dielectrics. For materials in common use, the capacitance does not change with voltage level. There are many parameters involved in selecting a commercial capacitor. Among the factors are voltage rating, shape, temperature stability, dielectric losses, and accuracy. The charge on the plates of a capacitor is the $D$ field times the area $A$ in square meters. $Q$ is therefore $\varepsilon_0\varepsilon_R EA$. The voltage across the plates is the $E$ field times the plate spacing $d$. Substituting $E = V/d$ in the equation for $Q$, it is easy to see that $C = Q/V = \varepsilon_R\varepsilon_0 A/d$, where the units for $A$ and $d$ are meters. A typical dielectric might have a relative dielectric constant of 10. For a plate area of 100 cm$^2$ and a spacing of 0.01 cm, the capacitance is 0.00885 $\mu$F.

## 2.5  CAPACITANCE IN SPACE

The space between a group of conductors can hold a complex electric field. This field is a function of the voltages on all the conductors. A specific portion of the field cannot be allocated to individual pairs of conductors, as a change in the potential for any one conductor modifies the entire field. Also, any change in field alters the charge distribution in all conductors that are embedded in the field. If there are electrical paths between conductors, charges will adjust to modify the field further. If there are no electrical paths, the charges will simply redistribute themselves on the various surfaces. In either case, the new field will reconfigure itself to store the least amount of field energy.

The concept of mutual capacitance is introduced to handle this complex situation. To measure a mutual capacitance, all conductors are grounded except one. This means that they are all connected to the reference potential, zero volts. It is convenient if a large conductive surface can be used as the zero of potential. The mutual capacitance between the ungrounded conductor and any other conductor is the ratio of charge on the grounded conductor to the voltage on the first conductor. This physical arrangement is shown in Figure 2.4. The self-capacitance of a conductor is the ratio of charge to voltage on that conductor with all other conductors grounded. The notation for mutual capacitance is $C_{12}$. The subscripts indicate the two conductors that are involved

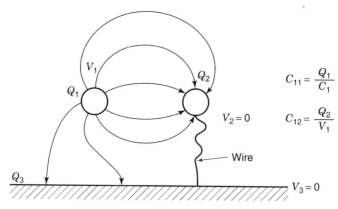

**FIGURE 2.4**  Mutual capacitance.

in the ratio of charge to voltage. A self-capacitance would be written $C_{11}$. A capacitor as a component has self-capacitance or simply, capacitance. Mutual capacitances are always negative, as the induced charge is always opposite in sign to the applied voltage. Like self-capacitance, mutual capacitance is a geometric quality.

In practice, an individual mutual capacitance is measured dynamically. This simply means that the applied voltage is varied and the resulting induced current flow is monitored. Because mutual capacitances are often very small, the techniques for measurement can be somewhat complicated. The difficulty arises because the measurement conductors modify the very geometry of the circuit being measured.

Field energy does not distribute itself so that it can be partitioned by mutual capacitances. The field at any one point has energy associated with every mutual capacitance. The stored energy can be calculated in terms of mutual capacitances, but the methods used would take us far from our main task. Suffice it to say that there is a complex field pattern between conductors in every circuit. The hope is that the field energy stored in components is much greater than the field energy stored between components. In high-speed circuits, mutual effects are an important consideration. For slower circuits the effect of mutual capacitance can be ignored. Exceptions can occur in active circuits. The transistors in an active circuit can multiply the effect of mutual capacitance. Circuit performance, and in some cases circuit stability, can depend on limiting certain mutual capacitances (see Section 5.14).

## 2.6  CURRENT FLOW IN CAPACITORS

Figure 2.5 shows a series circuit consisting of a capacitor, a switch, a series resistor, and a voltage source $V$. This is called a *series RC circuit*. At the moment the switch closes, there is no charge on the plates of the capacitor. This means that there can be no voltage across the capacitor terminals. At the moment of switch closure, all the battery voltage appears across the resistor. This voltage across the resistor means that there is a current flow equal to $V/R$. This current flow will begin to place charge on the capacitor plates. As the charge builds up on the plates, the $E$ field will increase between the plates.

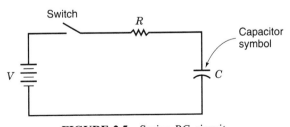

**FIGURE 2.5**  Series *RC* circuit.

This $E$ field implies a voltage across the plates of the capacitor. The sum of the voltages across the resistor and the capacitor must equal the impressed voltage. As the voltage across the plates of the capacitor rises, the voltage across the resistor drops reducing the current flow. Eventually after a long period of time, the charge stops flowing and the capacitor is said to be *fully charged*. The full charge is given by the formula $Q = CV$, where $C$ is in farads, $Q$ is in coulombs, and $V$ is in volts.

When the resistor is made smaller, a higher initial current will flow. This means that the capacitor charges more rapidly. A capacitor cannot be charged in zero time, as this requires infinite power. When the resistor is reduced to zero, there seems to be a conflict. At first glance the charging current would have to be infinite. On closer examination every battery has some internal resistance. The conductors of the circuit have some resistance, and even the plates of the capacitor have resistance. This means that in a practical circuit a capacitor is always charged through some finite series resistance. The charge time might be microseconds, but it can never be zero. There are other limitations to rapidly changing current flow, which will be discussed later.

Charge seems to flows around this simple circuit, but does it flow through the dielectric? The dielectric is an insulator, and yet somehow charge is moving onto one plate and off the other. The only answer that makes sense is to consider a changing $D$ field to be the equivalent of current. The more rapidly this field changes, the more current flows. When the capacitor is fully charged, the $D$ field is $\varepsilon_0 \varepsilon_R V / d$, where $\varepsilon_R$ is the relative dielectric constant. A changing $D$ field is, in effect, current flow in space. This current is called a *displacement current*. This idea makes it possible to consider a current flowing around the circuit, including the space between the plates.

The current flowing through a capacitor is proportional to the changing voltage that appears across the plates. The larger the capacitance, the greater the current. The faster the field changes, the greater the current. A capacitor opposes current flow, but it would be incorrect to say that it is a resistor. For a steady voltage the opposition to current flow is infinite. The opposition to current flow is discussed in Section 3.2.

## 2.7   *RC* TIME CONSTANT

After the switch is closed in Figure 2.5, the voltage across the capacitor begins to rise. The voltage plotted against time is described by an exponential curve where the exponent is $-t/RC$. The letter $t$ is time in seconds. This charging curve is shown in Figure 2.6. The product $RC$ has units of time when $R$ is in ohms and $C$ is in farads. When $t = RC$, the voltage reaches 63% of its final value $V$. This value of $t$ is called a *time constant*.

If the capacitor starts out with a voltage $V$ and a switch connects a resistor $R$ across its terminals, the voltage will begin to drop. This voltage also follows an exponential waveform. In one time constant the voltage drops to 63% of

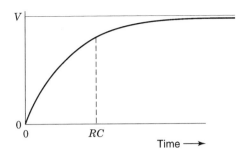

**FIGURE 2.6**   Exponential charging curve of a capacitor.

its initial value. It is easy to see that if two different voltages were alternately connected to the *RC* circuit, current would flow at all times.

If $R = 100,000$ Ω and $C$ is 10 $\mu$F, the time constant is 1 s. If the battery potential is 10 V, the voltage across the capacitor will rise to 6.3 V in 1 s. After the capacitor is fully charged, a 100,000-Ω resistor will discharge the capacitor to 3.7 V in 1 s. A 10,000-Ω resistor would discharge the capacitor to 3.7 V in 0.1 s. If the voltage were 100 V, the 10,000-Ω resistor would discharge the capacitor to 37 V in 0.1 s. The concept of time constant appears often in science.

The current that flows in the circuit of Figure 2.6 dissipates heat in the resistor. The potential energy that is finally stored in the capacitor is available to do external work at a later time. When the capacitor is discharged through a resistor, this potential energy is converted to heat. A capacitor stores field energy, it never dissipates energy.*

## 2.8   PROBLEMS

1. Two metallized sheets separated by an insulator are rolled into a cylinder to form a capacitor. The sheets are 3 cm wide by 1 m long. If the spacing between conductors is 0.03 cm and the insulator has a dielectric constant of 10, what is the capacitance of the capacitor?

2. Ten volts is placed across the terminals of the capacitor in problem 1. What charge is stored in the capacitor?

3. What is the $D$ field in problem 2?

4. A 10-V battery is in series with a switch, a resistor of 1000 Ω, and a 10-$\mu$F capacitor. At the moment the battery is connected to the *RC* circuit, what is the value of the current?

5. What is the *RC* time constant in problem 4?

---

*Practical capacitors can overheat from excess current flow. Manufacturers often list an equivalent series resistance so that the dissipation can be calculated.

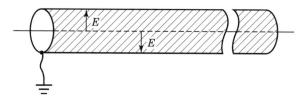

**FIGURE 2.7**   Shielded conductor.

**6.** A 10-$\mu$F capacitor is charged to 10 V. A 10,000-$\Omega$ resistor is placed across the capacitor terminals. What is the voltage after one time constant? What is the voltage after two time constants?

**7.** A resistor discharges a 10-$\mu$F capacitor to 37% of its initial voltage in 2 s. What is the resistor value?

**8.** Draw the current curve in problem 7. Assume an initial voltage of 10 V.

**9.** In problem 8, divide up the discharge current into 10 equal periods of time. Limit the maximum time to a period of one time constant. Estimate the energy dissipated in the resistor. How does this compare with the energy supplied from the capacitor? (*Hint:* The energy stored in the capacitor is $\frac{1}{2}CV^2$, where $C$ is in farads and $V$ is in volts.)

## 2.9 SHIELDS

The $E$ field around a conductor extends to all nearby conductors, and in theory it extends far out into space. It is posssible to limit the $E$ field around a conductor by changing the geometry. Figure 2.7 shows a grounded conductor surrounding the conductor in question. This conductor is called a *shield*. The field lines leave the center conductor and terminate on the inside surface of the shield. This geometry limits the $E$ field to the space between the two conductors. In this example there is no $E$ field or charge on the outside of the shield. If an external field did exist, it would terminate on the outer walls of the shield. This field would cause a charge distribution on the external surface. The external field cannot enter the space between the two conductors. The internal conductor is said to be shielded from external influences. Shields are not perfect, and small amounts of field can exit or enter directly through the shield. Fields can also enter at the ends of the shield, and this can be a significant problem. This problem is discussed in Section 6.14.

    The shield and its center conductor are called a *shielded cable*. If a voltage is placed between the shield and the center conductor, the $E$ field is confined to the inside of the cable. In most applications the shield is grounded, but this is not a requirement for there to be shielding.

    Shielding is not limited to cables. A conducting box can be a shield for a circuit. If the box surrounds the circuit, the $E$ field activity in the box is confined to the box and external $E$ fields cannot enter. A totally sealed box is impractical, as circuits must be ventilated and leads must be brought in and out of the box. Even if it is not perfect, shielding is a very important tool in electronic design. There are many techniques that get around the imperfect shield or box. Rapidly changing signals pose separate shielding problems. This is a topic for later chapters.

The shield of a typical cable is a conducting braid that surrounds a center conductor. This braid is made of many small tinned wires that allow for flexibility and an electrical connection at the ends. Braided cable has many small holes, and some of the internal $E$ field can find its way out of the cable. This $E$ field terminates on induced charges on external conductors. This leakage of the $E$ field is described correctly as a mutual capacitance. If the $E$ field varies, the induced charges must change. This changing charge implies current flow, which can be a source of a signal in an external circuit. The signal in the cable is said to be coupled to an external conductor through a mutual capacitance. This capacitance is also called a *leakage capacitance*. In most examples, leakage capacitances are on the order of a few picofarads per foot of cable.

## 2.10   MAGNETIC FIELD

The second electrical field found in nature is the magnetic field. Perhaps the most common magnetic field is the one that surrounds the earth. This field makes it possible to navigate through the use of a compass. Early experimenters found that a similar magnetic field was created near conductors that carried current. This means that there is both an $E$ field and a magnetic field whenever charges are moving along a conductor. Remember that charges will move along a conductor only if there is an internal electric field.

The magnetic field around a conductor carrying current is shown in Figure 2.8. This same field would exist if the current were a beam of electrons in a vacuum. The field resulting from a current is called an *H field*. This field is a force field, but this time the force is not on static charges. Evidence of the force exists when a steady current is passed through a piece of paper containing a sprinkling of iron filings. The filings tend to align themselves

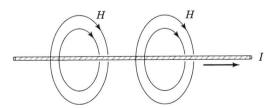

**FIGURE 2.8**   Magnetic field around a current.

with the field and form circles. A compass needle will also align itself with this magnetic field.

In Figure 2.8, the $H$ field continues out into the surrounding space. The field lines represent the alignment direction of a small compass. Note that the field lines or magnetic flux form closed circles. By convention the number of lines that are drawn is proportional to the current flow. These flux circles are spaced along the length of the current path. If the path is twice as long, the amount of flux is doubled.

The electrons that make up atoms have spin. In some materials the spin axis of electrons in adjacent atoms tends to line up in the same direction. It is this alignment that gives a material its magnetic properties. Each spinning electron generates a small magnetic field. The result of having many electrons spinning in the same direction is a magnetized material. Permanent magnets are materials that retain this alignment. Magnetic materials such as iron or steel are attracted to permanent magnets. Once in contact with a strong magnetic field these materials will also retain some magnetism. An example might be a magnetized iron nail or a steel screwdriver. When an iron bar or keeper is attracted to a permanent magnet, a considerable force may be required to pull the bar loose. This indicates that significant energy can be stored in a magnetic field.

There is no magnetic particle equivalent to the charge that can be used to measure the magnetic field force and direction. If such a particle did exist, it would be called a *monopole* (a single magnetic pole). Physics books use a hypothetical element of current to characterize the magnetic field. This is used because the forces between two elements of current are well understood. The current element must be small enough so that it does not modify the field it is measuring. Just like the $E$ field, the magnetic field exerts a force at a distance. Where the flux lines are close together, the magnetic field intensity is greatest.

## 2.11 SOLENOIDS

When the current path consists of many turns of wire along a cylinder, the $H$-field intensity in the center of the turns is proportional to the number of turns. This conductor geometry is called a *solenoid*. Note that the $H$-field lines are no longer circles. Flux lines are still loops that close on themselves. A solenoid and its associated $H$ field is shown in Figure 2.9. The $H$-field intensity is greatest and nearly constant everywhere inside the solenoid. Outside the solenoid the $H$-field intensity drops off significantly.

## 2.12 AMPÈRE'S LAW

In the $E$-field case, a force exerted over a distance represented work and the work per unit charge was defined as voltage. In the magnetic case, the force

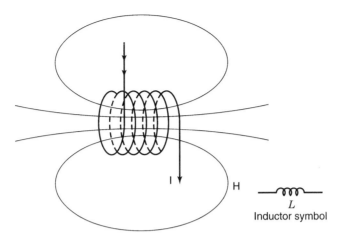

**FIGURE 2.9**   Simple solenoid and its associated $H$ field.

exerted over a distance also represents work, but here the work must be done on a unit segment of current or on that imaginary monopole. The product of $H$-field intensity × distance turns out to be a measure of current. In the $E$-field case the $E$-field intensity times the distance between the plates yielded the voltage across the plates. In the $H$-field case the $H$-field intensity times the distance around the flux path yields the current that creates the flux.

An external circuit supplies the current that creates the magnetic field. In ordinary circuits, energy must be dissipated continuously to sustain a steady current. This energy is lost in the resistance of conductors that make up the current path. The positive work done in moving a current element through a magnetic field reduces the amount of energy that must be supplied by the external circuit. In an ideal circuit where the resistances are zero, the positive work done on the current element would increase the current in the loop.

In Figure 2.8, the $H$-field intensity is constant along any flux path. The product of $H$ × the total path length is a measure of current in the nearby conductor. This fact is known as *Ampère's law*. The path length is $2\pi r$, and therefore $2\pi r H = I$. This means that $H = I/2\pi r$, where $r$ is in meters and $I$ is current in amperes. The units of $H$ are thus amperes per meter. Compare this with the units for the $E$ field, which are volts per meter.

Ampère's law can be applied to the solenoid. In this case the $H$-field intensity changes along the flux path. This can be seen in Figure 2.9 where the $H$-field lines are not uniformly spaced. In the $E$-field case, the product of $E$ × distance had to be considered over short distances. The $H$ field must be treated in the same way. The product of $H$ × distance must be made over short distances, where $H$ is nearly constant. If these products are summed around the flux loop, the answer will be the current enclosed. In the case of a solenoid with $n$ turns, the answer will be $nI$, where $I$ is the current in the solenoid. In equation form, $H = nI/2\pi r$, where $I$ is in amperes and $r$ is in meters.

## 2.13 PROBLEMS

1. A conductor carries a 3-A steady current. What is the $H$-field intensity 10 cm away from the center of the conductor?

2. One hundred turns of wire are evenly spaced around a toroid. What is the $H$ field in the center of the toroid if the radius is 5 cm and the current is 0.1 A?

3. Solve problem 2 where the radius of the toroid is 10 cm.

4. Is the $H$ field in the toroid different if the turns are square?

5. A conductor 2 cm in diameter carries a dc current of 10 A. What is the $H$ field inside the conductor at a depth of 0.5 cm? What is the $H$ field at the center of the conductor?

## 2.14 MAGNETIC CIRCUIT

Most magnetic applications require the use of specially alloyed magnetic materials where iron is usually one of the components. These alloys play a role in magnetics just as a dielectric plays a role in electrostatics. In the $E$-field case, the dielectric reduced the intensity of the $E$ field in the dielectric. In the magnetic field case the magnetic material reduces the intensity of the $H$ field in the magnetic material. The ability to reduce the $H$ field is called *relative permeability*. If the $H$ field is reduced by a factor of 100, the relative permeability of the material is 100.

Ampère's law still holds when magnetic materials are introduced in the flux path. If the current is held constant, the $H$ field adjusts so that it is reduced in the magnetic material and increased in the remaining air space. If the magnetic material forms a loop with a small air gap, the $H$ field will be very intense in the air gap. The iron and the gap form what is known as a *magnetic circuit*. A typical magnetic circuit is shown in Figure 2.10. In this example the length of the gap is 0.1 cm, the length of the magnetic path is 10 cm, the relative permeability is 1000, and the current is 0.1 A. The flux path in the iron is 0.099 m and in the gap it is 0.0001 m long. The $H$ field in the iron is $1/1000$ the $H$ field in the air gap. Ampère's law requires that the $H$-field intensity times the path length must add up to the total current. This means that $(H/1000) \times 0.099 + H \times 0.0001 = 0.1$ A. The value for $H$ is 502.5 A/m. This is the field strength in the air gap. The $H$-field intensity in the iron is $1/1000$ of this value, or 0.5025 A/m.

## 2.15 INDUCTION OR $B$ FIELD

The $H$ field is reduced in a material with permeability. This parallels the $E$ field case where the $E$ field is reduced in a dielectric. It is convenient to

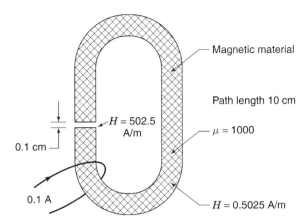

**FIGURE 2.10**   Magnetic circuit.

introduce a second magnetic field that is not changed by permeability. This field is called the *B field* or *induction field*. The ratio of the *B* field to the *H* field is the permeability of the material. The units of *H* are amperes per meter and the units for *B* are teslas. Another common unit for the *B* field is the gauss (G); 10,000 gauss is equal to 1 tesla. In equation form, $B = \mu_R \mu_0 H$, where $\mu_R$ is the relative permeability and $\mu_0$ is the permeability of free space. The value of $\mu_0$ is $4\pi \times 10^{-7}$. This constant is needed to convert amperes per meter to teslas. Note that $\mu_R$ for air is unity.

   The *B* field is the true force field on moving charges. This force is perpendicular to both the *B*-field direction and the direction of the moving charge. When charged particles from the sun get near the earth, they encounter the earth's magnetic field. The *B* field causes the charges to spiral in the magnetic field as they near the earth. This spiraling releases energy in the form of radiation. This is familiar to most people as the *aurora borealis*. This colorful display is a polar phenomenon, as this is where the magnetic field is most intense.

   A strong duality exists between the electric field and the magnetic field. When the *D* field changes in space, it is equivalent to current flow. When the *B* field changes in space, it turns out to be equivalent to voltage in space. This can be observed by placing a loop of wire in a changing magnetic field. The voltage that is induced in this loop is proportional to the amount of *B* flux that crosses the loop and how rapidly this flux is changing.* A larger loop area captures more flux, and this increases the voltage across the terminals of the loop. If several turns couple to the same flux, the induced voltage will be proportional to the number of turns. This circuit is shown in Figure 2.11, where two turns of wire couple to the changing *B*-field flux. The changing

---

*The term *flux* is used interchangeably with the term *field*. It is sometimes a help to visualize a portion of the field as lines of flux crossing a given area in space.

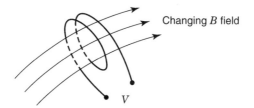

**FIGURE 2.11**   Voltage coupling from a changing $B$ field.

magnetic field forces charges to move to the surface of the added coil. The coil end that is positive has positive charges and the negative end has negative charges. These charges cannot move unless a circuit path is provided across the coil ends. The voltage that is coupled to the loop of wire means that there is an $E$ field whenever there is a changing $B$ field.

The voltage induced into a loop by a magnetic field is given by the rate of change of flux expressed in webers per second. This is known as *Faraday's law*. To obtain the flux in webers, the $B$ field in teslas must be multiplied by the cross-sectional area of the loop in square meters. To obtain the $B$ field in teslas, the $H$ field must be divided by the permeability of free space and the relative permeability. Faraday's law in equation form is $V = n \times$ (the rate of change of magnetic induction flux per second), where the magnetic flux is in units of maxwells, $V$ is in volts, and $n$ is the number of turns.

It is important to recognize that Faraday's law works two ways. If there is a changing magnetic flux crossing an open conductive loop, a voltage can be sensed at the ends of the loop. Conversely, if a voltage is placed across a conductive loop, there must be a corresponding changing flux crossing the loop. This changing flux implies current flow supplied by the voltage source.

## 2.16   MAGNETIC CIRCUIT WITHOUT A GAP

Figure 2.12 shows a simple transformer that has a core without a gap. The magnetizing properties of this material are considered ideal. This means that $B$ and $H$ are always proportional to each other. Practical magnetic materials have hysteresis, which means that there is nonlinear relationship between the $H$ and $B$ fields. The current required to establish the $H$ field is called a *magnetizing current*. If the magnetic material were ideal (an infinite permeability), the magnetizing current would be zero.

Two coils are wrapped around the magnetic path so that any magnetic flux in the path threads through both coils. When a steady voltage is impressed across the first coil, the flux that threads through that coil must change at a given rate. A steady voltage implies a constantly changing flux and a constantly changing flux, implies a steady voltage. This statement is Faraday's law. This rate of change in magnetic flux requires a corresponding rate of change in the $B$ field. This rate of change in the $B$ field requires a correspond-

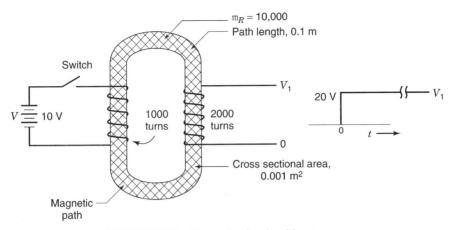

**FIGURE 2.12**  Magnetic circuit without a gap.

ing rate of change in the *H* field. This further requires a rate of change in the current flowing in the coil. The *B*-field intensity in the magnetic path will increase until the core saturates (the iron cannot handle any more flux). At this point the permeability drops and the current supporting the *H* field will increase sharply. To be practical, the *B* field must be kept within bounds. This means that the impressed voltage can last only a limited period of time. In this magnetic circuit the only function of the magnetizing current is to establish the *B* field.

Here is an example of how this magnetic circuit works. The magnetic path length in Figure 2.12 is 0.1 m long. The maximum *B*-field intensity in this material might be 1 tesla (T). Assume the following parameters: The relative permeability is 10,000, the voltage impressed across the first coil is 10 V, and the coil has 1000 turns. The voltage per turn is 0.01 V. Using Faraday's law, this voltage requires that the flux coupling to each turn must change at 0.01 weber (Wb) per second. (One turn and 1 V supports 1 Wb/s.) The *B*-field intensity is simply this flux divided by the cross-sectional area. If the cross-sectional area is 0.001 m², the *B* field must change at the rate of 10 T/s. To find the *H* field, the *B* field must be divided by both the permeability of free space ($4\pi \times 10^{-7}$) and the relative permeability of the iron (10,000). The rate of change of the *H* field is $10^4/4\pi$ amperes per meter per second. Because the path length is 0.1 m and the number of turns is 1000, the current in the coil must change at $1/4\pi$ amperes per second. In 0.01 s the current in the coil will go from zero to 0.796 mA, a small magnetizing current. Remember that the *B* field is changing at 10 T/s. At the end of 0.01 s the *B* field has reached an intensity of 0.1 T. If the voltage is left connected to the coil for a full 0.1 s, the *B* field will increase to 1 T and the magnetic material will just saturate. At this point the *H* field has also increased and the magnetizing current is up to 7.96 mA. If the magnetic material saturates and the permeability drops to

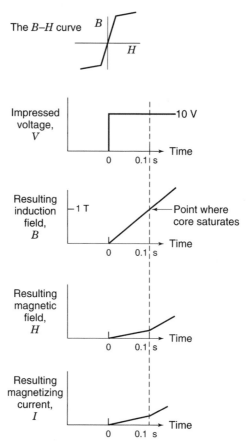

**FIGURE 2.13** Waveforms associated with impressing a voltage on a coil in a magnetic circuit.

100, the $H$ field must increase by another factor of 100. This means that the magnetizing current will rise. It is obvious that the permeability must remain high if the magnetizing current is to remain low. See Figure 2.13 for the waveforms in this example.

## 2.17 MAGNETIC CIRCUIT WITH A GAP

Figure 2.14 shows the previous magnetic circuit with a gap of 0.1 cm. The total path length is 0.1 m. The maximum $B$ field before core saturation is again 1 T. A voltage of 10 V is applied across 1000 turns, as before. The magnetic material has a relative permeability of 10,000. Faraday's law requires that the flux crossing each coil must change at 0.01 Wb/s. The $B$-field intensity is simply the flux change divided by the cross-sectional area of the

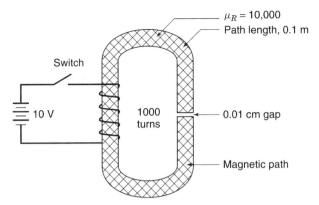

**FIGURE 2.14** Magnetic circuit with a gap.

magnetic material. As before, this area is 0.001 m² and the *B*-field intensity must change at 10 T/s. The *H* field intensity in the air gap is the *B* field divided by the permeability of free space or $4\pi \times 10^{-7}$. The *H*-field intensity in the iron is reduced further by the relative permeability. The *H* field in the air gap must change at $7.96 \times 10^6$ A/m per second and the *H* field in the iron must change at $7.96 \times 10^2$ A/m per second. The current flowing in one turn can be calculated by multiplying by the proper magnetic path length. The changing current required to support the *H* field in the gap is $7.96 \times 10^6$ times 0.001 m, or $7.96 \times 10^3$ A/s. In the iron the path length is 0.1 m and the magnetizing current changes at 79.6 A/s. The total requirement is the sum of the two changing currents, or 8039 A/s per turn. Because there are 1000 turns, the current in the coil changes at 8.039 A/s. In 0.01 s the current rises to 0.080 A. Note how the presence of the gap has increased the required current. This added current is storing field energy in the gap.

## 2.18 TRANSFORMER ACTION

The second coil in Figure 2.12 surrounds the magnetic circuit with a different number of turns. The voltage per turn on this coil is identical to the voltage per turn on coil 1. If the number of turns on the second coil is double, the voltage is double. Using the previous example, if the second coil has 2000 turns, the steady voltage across the coil would be 20 V. The two coils and the core would be called a *step-up transformer*. If the second coil had 500 turns, the terminal voltage would be 5 V and this would be called a *step-down transformer*.

In our ideal transformer the resistance of the coils is considered zero. What happens when a resistor is placed across the ends of the second coil? A current must flow that obeys Ohm's law. Assume that coil 2 has 2000 turns. If 10 V is impressed on coil 1, there will be 20 V across coil 2. If there is a 2000-Ω resistor across coil 2, a current of 10 mA will flow. This current flows steadily and does not increase in time like the magnetizing current. This current causes an $H$ field in the iron that seems to modify the ratio of $B$-field to $H$-field intensity. This does not happen. A current flows in coil 1 that exactly cancels this added $H$ field. If 10 mA flows in coil 2 with 2000 turns, a current of 20 mA must flow in coil 1, as it only has 1000 turns. In effect, there is a balance of ampere turns (At). Coil 2 has 10 mA in 2000 turns, or 20 At. This is balanced by coil 1, which has 20 mA in 1000 turns, or the same 20 At.

Coil 1 is called the *primary coil* and is connected to the source of voltage. Coil 2 is called a *secondary coil* and has a voltage that is determined by the number of turns and the voltage per turn impressed on the primary coil. Of course, there can be more than two coils wound around a magnetic circuit. The ampere turns of all secondary coils must add up to the ampere turns on the primary coil. The magnetizing current is independent of this transformer action.

Obviously, the steady voltage applied to the primary coil must eventually be removed or the magnetic material will saturate. If the voltage is reversed before core saturation, the $B$ field then reverses direction. In our example, the $B$ field reached 0.1 T in 0.01 s. If the voltage is reversed, the $B$ field will return to 0 in another 0.01 s and then continue to $-0.1$ T in 0.02 s. If the voltage is reversed every 0.02 s, the $B$ field will swing back and forth between $+0.1$ and $-0.1$ T. This reversal of voltage is called a *square wave*. One full cycle or duty cycle occurs every 0.04 s. This is also known as a 25-hertz (Hz) square wave (the unit hertz means cycles per second). The repetition rate is also called *frequency*. The reciprocal of duty cycle time is frequency in units of hertz.

The square wave of voltage is only one specific voltage waveform. The secondary voltages in an ideal transformer will follow exactly the same waveform on all secondary coils. This duplication of waveform occurs in a transformer where the magnetizing current is small and the coil resistances are very low. There are many factors that influence the performance of practical transformers, but it is important to understand the ideal case first.

## 2.19  MAGNETIC FIELD ENERGY

A magnetic field stores energy. The field under consideration can be created by moving an ideal current element into a closed loop. To perform this experiment, consider a conductive loop with zero resistance so that the current that flows cannot dissipate energy in heat. The magnetic flux that is associated with this current loop exerts a force on an external current element. Every time a current element is merged with the conductor, more current is added to the conducting loop, which increases the field strength.

The work done on moving a current element in a magnetic field is stored in that magnetic field. For a straight piece of wire, field energy must be stored along the entire current path. If the path is 10 units long, 10 current elements must be moved to increase the current by one unit. If the path length is 20 units long, 20 current elements must be moved to increase the $H$-field intensity to the same level.

The $H$ field around a conducting loop is proportional to the current in that loop. The work done in adding an element of current is proportional to the $B$ field (the force field) and the increment of $H$ field that is added. The energy stored in a small volume $V$ where $B$ and $H$ are constant is $\frac{1}{2}HBV$. When $B$ is in teslas, $H$ is in amperes per meter, the volume is in cubic meters, and the energy units are joules.

The energy stored in the magnetic circuit of Figure 2.14 can now be considered. In the iron the $H$ field is reduced by 10,000 and the path length is roughly 0.1 m. In the gap the $H$ field is not reduced and the path length is 0.001 m. The ratio of the energy stored in the gap to the energy in the iron is 100. From this example it is easy to see that 99% of the magnetic energy is stored in space (the gap) not in the iron. The iron is there to confine the flux so that energy can be stored in the gap. Components that store magnetic field energy are called *inductors*. The energy in the gap of Figure 2.13 is left as an exercise.

## 2.20  INDUCTORS

Inductors are components that store magnetic field energy. The unit of inductance is the henry (H). In most applications the henry is a large unit and it is common to see component values of microhenries ($\mu$H) and millihenries (mH). The letter $L$ is used in circuit diagrams to designate an inductor. The statement $L = 3$ mH is read as an inductor having an inductance of 3 millihenries. The electrical symbol for the inductor is shown in Figure 2.9.

Inductance is defined as the $B$ flux generated per unit of current. It is a geometric property of conductors and their magnetic circuits. A straight piece of wire has inductance because a current flowing in the wire has an associated magnetic field. The solenoid in Figure 2.9 is a very common form of inductor where wire is wound over a cylindrical core made of a magnetic material. The inductance is varied by changing the core material, core size, and number of turns.

The measurement of inductance is usually made by noting the voltage when the current changes in the inductor. This voltage is always in the direction to oppose the current flow. The presence of this voltage is known as *Lenz's law*. If a current is increasing at 1 A/s, the resulting voltage is 1 V when the inductance is 1 H. The inverse is also true. If 1 V is applied across an inductor of 1 H, the current in the inductor will increase at 1 A/s. This parallels the function of a capacitor. If a current of 1 A flows into the plates of a 1-F capacitor, the voltage will rise at 1 V/s. This is another example of the duality between electric and magnetic phenomena.

A magnetizing current flows when a voltage is impressed across the terminals of an unloaded transformer. Faraday's law requires a changing $B$ field.

A current must flow to establish the corresponding $H$ field. This increasing current means that an unloaded transformer looks like an inductor. Because there is no intentional air gap, this inductance is not suited for storing magnetic field energy. To illustrate this point, consider a nearly ideal transformer. Here the magnetizing current is near zero, which means that the magnetizing inductance is very large. Without current flow there can be no energy storage.

Inductors, capacitors, and resistors are the basic elements of lumped-parameter circuits. An analysis of a circuit does not directly treat the fact that the capacitors and inductors store field energy and that these ideal elements never dissipate energy. A circuit diagram is a simplification of what actually takes place. There is field energy stored in and around each component and the interconnections carry field energy to the components in the right sequence. This movement of field energy on conductors is an important subject and is treated in subsequent sections.

The energy stored in an inductor is proportional to the square of the current flowing in the inductor. This follows from the ideas discussed in Section 2.19. The magnetic field energy in any small volume of space is proportional to $H$ squared. But $H$ is proportional to current flow. This is the reason that the energy stored in an inductor is proportional to the square of the current. The proportionality factor is the inductance. The equation for energy storage is thus $E = \frac{1}{2}LI^2$, where $E$ is in joules, $L$ is in henries, and $I$ is in amperes. This should be compared with the energy stored in a capacitor, which is $\frac{1}{2}CV^2$.

## 2.21  $L/R$ TIME CONSTANT

Energy stored in an inductor cannot be changed instantly, as this takes infinite power. This means that the current in an inductor cannot change instantly. Consider the simple circuit shown in Figure 2.15. At the time the switch is closed, the magnetic field energy is zero. This means the current in the inductor is zero. Without current flow there is no voltage drop across the

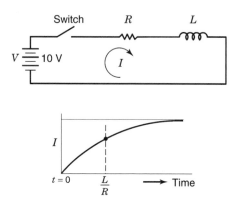

**FIGURE 2.15**   Simple circuit containing an inductor.

resistor. This means that initially the full battery voltage appears across the inductor. Faraday's law requires that amperes per second times the inductance must equal this battery voltage. As the current builds up, the voltage drop across the resistor increases. The sum of the voltages across the inductor and resistor must equal the impressed battery voltage. As time proceeds, the voltage across the resistor increases and the voltage across the inductor falls. After a length of time the full voltage appears across the resistor and there is no voltage across the inductor. The current curve is an exponential function of time, as shown in Figure 2.14. The final current value is simply $V/R$. Note that the resistor–capacitor voltage curve shown in Figure 2.5 is exactly the same as this current curve. The voltage across the inductor is said to fall off exponentially with time.

The current in Figure 2.14 rises to 63% in one time constant. The ratio $L/R$ has units of time and is a time constant. For a battery voltage of 10 V, an inductor of 1 H, and a resistor of 1 $\Omega$, the current rises to 6.3 A in 1 s. If $L$ is 1 H and $R$ is 10 $\Omega$, the current will rise to 0.63 A in 0.1 second. Consider an ideal inductor with no internal resistance and an external resistance that approaches 0 $\Omega$. The time constant $L/R$ is obviously very large. This simply means that once a current is established, it continues to flow and never diminishes. This situation actually exists if the conductors in the inductor are superconductive. This continuous current flow represents stored magnetic field energy, analogous to a lossless flywheel storing kinetic energy. A capacitor storing electric field energy is analogous to the potential energy of a raised weight. Under the right circumstances the stored energy in either field can be released to do external work.

The analysis of the $LR$ circuit in Figure 2.14 assumes that the inductor core material does not saturate. A steady flow of current in turns around a core can saturate that core, which basically sets the inductance to near zero. Of course, an air-core inductor cannot saturate. In circuits with currents limited to a few milliamperes, saturation is not likely to take place. Inductors used in power circuit filters can be saturated from line current, making the component useless.

## 2.22   MUTUAL INDUCTANCE

The ratio of current in a first circuit to the magnetic flux coupled to a second circuit is called a *mutual inductance*. The circuit designator is $L_{12}$, where the subscripts indicate the two circuits that are considered. A self-inductance equals the flux in a circuit generated by the current in that same circuit. The circuit designator is $L_{11}$ or simply $L$. Mutual inductance is just like mutual capacitance: It is a function of circuit geometry only. A current does not need to flow for there to be mutual inductance.

Mutual inductance can be of either polarity. When the flux is changing, the voltage induced by Faraday's law can be of either polarity, depending on the orientation of the coupling loop. Mutual inductance can be the mechanism that allows a nearby current to induce an unwanted signal into a circuit. Unwanted coupling can be reduced by a suitable change in circuit geometry. The interfering circuit can be moved or designed to limit the extent of any external flux. The sensitive circuit can be moved or the coupling loop area limited.

## 2.23 PROBLEMS

1. A dc voltage of 2 V is impressed across an inductor of 1 mH. How much time elapses before the current reaches 10 mA?

2. In problem 1, how much energy is stored in the inductor?

3. In a transformer the magnetic path is 20 cm long. The relative permeability of the magnetic material is 20,000. What is the $H$ field if the magnetizing current is 20 mA flowing in 500 turns?

4. What is the $B$ field in problem 3?

5. If the core material in problem 3 is has an area of 5 cm², what is the flux level in webers?

6. Twenty volts is impressed across the 500 turns of problem 3. How much time must elapse before the magnetizing current reaches 20 mA? 30 mA?

7. The operating flux level for the magnetic material is 0.5 T. What is the lowest-frequency square wave that can be used across the coil? Assume a square wave that is 10 V peak to peak. The square wave is symmetrical about 0 V.

8. How much energy is stored in an air gap 0.1 cm long with an area of 15 cm²? Assume a $B$ field of 1 T. How much energy is stored if the gap is 0.05 cm long?

9. In problem 8, what is the $H$ field in the gap? How much current must flow in 100 turns to establish this $H$ field? Assume a negligible $H$ field in the magnetic material.

## 2.24 REVIEW

Capacitors store electric field energy. The use of dielectrics increases the charge that can be stored in a given geometry. When the $D$ field is changing in a capacitor it is equivalent to current flow. This current is called a *displacement current*. This same concept is valid in space, where the $E$ field is not confined to a capacitor. Shields provide a geometry that that can contain an electric field. A shield might be a conductive braid around a conductor or a metal box around a circuit. Shields can limit the entry of external fields or keep local fields contained.

Magnetic fields store energy in space. Magnetic materials can be used to focus where this energy is stored. Inductors are components that are designed specifically to store magnetic field energy. Transformer action can use magnetic materials to transfer a varying voltage from one coil to another by using a changing magnetic field. A changing magnetic field is equivalent to a voltage in space. Faraday's law requires that the rate of change of $B$ flux produces a voltage in any open conductive loop. Ampère's law provides a way to determine the intensity of the $H$ fields that are produced by current flow. The relationship between the $B$ and $H$ magnetic fields involves the relative permeability and the permittivity of free space.

There is an analogy between field energy storage and mechanical energy storage. A magnetic field can be considered the storage of kinetic energy (a current is a moving charge). The parallel might be a frictionless flywheel. An electric field can be considered the storage of potential energy. The analogy might be the potential energy of a weight that has been raised in height. In either case the energy is stored and is available to do work at a later time. Stored energy can only be dissipated in heat, sound, light, mechanical work, or radiation. Energy is never lost in an ideal capacitor or inductor. An ideal transformer transfers energy and never dissipates energy. The transformer allows the $E$ field to be increased at the expense of $H$ field, and vice versa.

# 3 Utility Power and Circuit Concepts

## 3.1 SINE WAVES

Sine-wave behavior is found in everyday phenomena. A pendulum moves back and forth with a sinusoidal motion. A weight hanging on a spring moves up and down with a sinusoidal motion. The length of the day varies over a year in a sinusoidal manner. The height of a point on a turning wheel varies sinusoidally. A conducting coil rotating in a magnetic field generates a sinusoidal voltage. This is the basis of power generation used throughout the world. In all these examples, time is involved in a cyclic motion. The number of cycles per second is called *frequency* and has units of hertz. The standard power frequency in North America is 60 Hz.

Circuits are available that generate sine-wave signals. The source of energy for these circuits can be a battery or the utility power. These sine-wave sources supply a varying voltage that remains constant in character as the load changes. These voltage sources are said to have a low source impedance. A typical laboratory sine-wave generator can generate signals that can be varied from a few hertz to many megahertz.

When a sinusoidal voltage is impressed across the terminals of a capacitor, a sinusoidal current flows. Similarly, a sinusoidal voltage impressed across an inductor causes a sinusoidal current to flow. This is the only voltage waveform that copies itself as a current waveform in capacitors and inductors. A sinusoidal voltage impressed on any configuration of linear passive components ($R$, $L$ or $C$) will cause sinusoidal voltages and currents to appear in the entire circuit. For this reason sine waves form the basis for most circuit analysis. It is possible to introduce other waveforms, but the analysis is very mathematical. Circuits are often described by how they respond to sine waves of various frequencies. A great deal of time is spent in education on the subject of frequency analysis. The inverse problem, called *network synthesis*, is taught as a more advanced topic. This subject treats the problem of finding a circuit that matches a required frequency response.

A range of frequencies is called a *band* or a *bandwidth*. The human ear can hear sounds where the air pressure varies sinusoidally in the band from about 19 Hz to about 18 kHz. AM radio signals are broadcast over the band 0.5 to 1.8 MHz. A video pattern might have a bandwidth that includes

50 Hz and 4.5 MHz. The world of electronics uses sine waves to describe and measure performance. Sine waves are not the only waveform that can be used in measuring circuit performance. Square waves can be used to measure response time and indicate if there are any instabilities.

## 3.2   REACTANCE AND IMPEDANCE

When a sinusoidal voltage is impressed across a capacitor, the sinusoidal current that flows is maximum when the voltage is changing most rapidly. The current is maximum when the voltage waveform goes through 0 V. If a sinusoidal voltage is connected to a capacitor at a positive-going zero crossing, the initial current in the capacitor is at a positive maximum. When the voltage is at a maximum, it is no longer increasing or decreasing, and this is where the current is zero. The current waveform is thus shifted in time so that the peak of current occurs before the peak of voltage. The current is said to *lead* the voltage.

When a sinusoidal current flows in an inductor, the voltage that appears across the inductor terminals is also sinusoidal. If a sinusoidal voltage is connected to an inductor at its peak value, the current starts out at zero. As the voltage drops (it is still positive), the current increases. When the voltage reaches zero, the current is at its peak. When the voltage reaches its maximum negative value, the current is again zero. The current waveform is thus shifted in time from the voltage. The current is said to *lag* the voltage.

The maximum slope of a voltage sine wave is $2\pi fV$, where $V$ is the peak voltage and $f$ is the frequency. The current that flows in a capacitor is proportional to the capacitance and to the slope of this voltage. The ratio of peak voltage to peak current is thus $1/2\pi fC$. This is known as *capacitive reactance*, designated by the letter $X_C$. The unit of capacitive reactance is the ohm.

A sinusoidal voltage impressed across the terminals of an inductor causes a sinusoidal current to flow. This current flow decreases if the inductance is made greater. If the voltage changes more rapidly, the current is reduced. The opposition to current flow is thus proportional to inductance and to the rate of change of current. The ratio of peak voltage to peak current is $2\pi fL$. This expression is called *inductive reactance* and is designated by the letter $X_L$. The unit of inductive reactance is the ohm.

When a resistor and an inductor are placed in series, the opposition to current flow is not the sum of the reactance and the resistance. This is because the voltage across the resistance peaks with the current, but the voltage across the inductance at that moment is zero. At any instant in time the sum of the voltages across the inductor and the resistor must equal the impressed voltage. Since the voltage maximums do not occur at the same time, the peak values cannot be summed directly.

The simplest way to understand what is happening is to consider a pointer rotating in a circle. Let the pointer length represent the current that is flowing.

When the pointer is straight up, the current is maximum. The height above the horizontal represents the intermediate values of current. When the pointer is pointing to the right, the current is zero. Convention requires that the pointer move counterclockwise around the circle once every cycle of the sine wave. The voltage across the resistor tracks with this current. A second pointer representing this voltage rotates at the same rate and is horizontal at the same point in time. The voltage across the inductor leads the current. When the current is zero, the voltage across the inductor is maximum. This pointer is straight up when the current pointer is pointing to the right. This $90°$ relationship continues as the pointers rotate.

Imagine a strobe light that illuminates the pointers once per cycle. The light can be timed so that the current pointer is set to point to the right as a reference. The pointers can be used to determine the sum of the voltages across the resistor and the inductor. Construct a rectangle from the two voltage pointers. The diagonal of this rectangle is a pointer that represents the sum of the two vertical voltage components. The diagonal length is simply $\sqrt{(IR)^2 + (IX_L)^2}$. The three pointers are shown in Figure 3.1. The ratio between voltage and current is called an *impedance* and is equal to $\sqrt{R^2 + X_L^2}$.

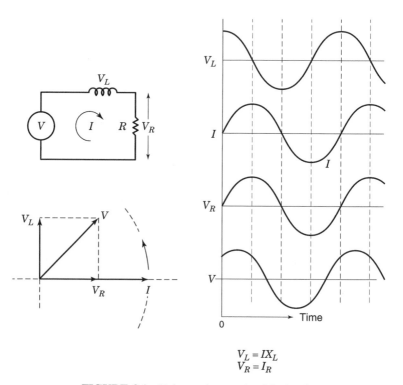

$$V_L = IX_L$$
$$V_R = I_R$$

**FIGURE 3.1**   Voltages in a series *RL* circuit.

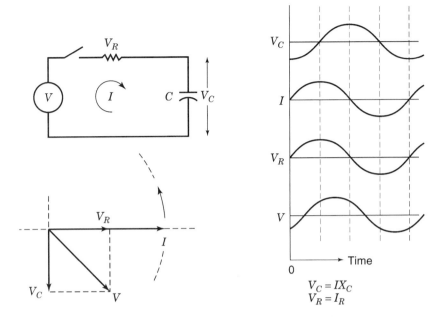

**FIGURE 3.2**   Voltages in a series *RC* circuit.

Consider current flowing in a series capacitor and a resistor (Figure 3.2). The same pointer idea can be used. The voltage pointer for the resistor follows the current pointer. The voltage pointer for the capacitor lags the current pointer. When the current pointer is horizontal, the voltage pointer for the capacitor points straight down. This is in the direction opposite to the pointer for the inductor case. The sum of the voltages is represented by the length of the diagonal of a rectangle formed by the two voltage pointers. The voltage across the resistor is $IR$ and the voltage across the inductor is $IX_L$. The length of the diagonal is $\sqrt{(IR)^2 + (IX_L)^2}$. The ratio of voltage to current is $\sqrt{R^2 + X_L^2}$. This is also known as an *impedance*, with units of ohms.

Energy is stored in the inductor or capacitor twice per cycle. The energy storage is zero in the inductor when the current is zero. The energy storage in the capacitor is zero when the voltage across the capacitor is zero. Energy is dissipated in the resistor except when the current is zero.

When a sinusoidal voltage is impressed across a series resistor–capacitor circuit as in Figure 3.2, the voltage $V_C$ across the capacitor decreases as the frequency is increased. This is known as a simple *low-pass RC filter*. When the reactance of the capacitor is much smaller than the resistance, the voltage across the capacitor will lag the applied voltage by nearly 90°. This phase shift is always associated with a proportional *attenuation* slope. (The attenuation is directly proportional to frequency.) At the frequency where the resistance equals the reactance, the attenuation factor is $\sqrt{2}/2$.

When a sinusoidal voltage is impressed across a series inductor–resistor circuit as in Figure 3.1, the voltage across the resistor decreases as the frequency is increased. This known as a simple *RL* filter. When the reactance of the inductor is greater than the resistance, the voltage across the resistor will lag the applied voltage by nearly 90°. Again the attenuation slope is proportional to frequency. At the frequency where the reactance equals the resistance, the attenuation factor is $\sqrt{2}/2$.

Passive filters made up of groups of inductors and capacitors can be used to attenuate bands of frequencies selectively. This is the subject of filter theory taught in advanced circuit theory classes. Filters can be designed to have low-pass, high-pass, band-pass, or band-reject character. This same selectivity in terms of frequency can often be achieved using active components (integrated circuit amplifiers) or digitally using a sampling algorithm.

## 3.3  PROBLEMS

1. What is the reactance of a 1-$\mu$F capacitor at 100 Hz, 1 kHz, and 10 kHz?

2. What is the reactance of a 1-H inductor at 100 Hz, 1 kHz, and 10 kHz?

3. A series circuit consists of a 159-$\Omega$ resistor and a 1-$\mu$F capacitor. What is the impedance of this circuit at 100 Hz, 1 kHz, and 10 kHz?

4. A series circuit consists of a 1-H inductor and a 6.28-k$\Omega$ resistor. What is the impedance at 100 Hz, 1 kHz, and 10 kHz?

5. A resistor of 5000 $\Omega$ is in parallel with a 0.02-$\mu$F capacitor. A voltage of 10 V peak at 1 kHz is placed across the two components. What is the peak current in each component? (*Hint*: The total peak current can be found by finding the length of a diagonal where the current lengths are two sides of a rectangle.) What is this peak current? Use this peak current to find the impedance of the circuit.

6. In problem 4, what is the impedance of this parallel circuit at 5 kHz?

7. A resistor of 10,000 $\Omega$ is in parallel with a 20-mH inductor. A voltage of 20 V peak at 2 kHz is placed across the parallel circuit. What is the peak current in each component? Use the rectangle rule to fine the total peak current. What is the impedance of this parallel circuit?

8. An inductor of 3 mH is in parallel with an inductor of 1 mH. Use the sum of current idea to determine the parallel reactance.

## 3.4  RESONANCE

The voltages across the inductor, capacitor, and resistor in Figure 3.3 must all add up at any instant to the impressed voltage. The same pointer scheme used for the *RL* and *RC* circuits can be used for this *RLC* circuit. The current pointer

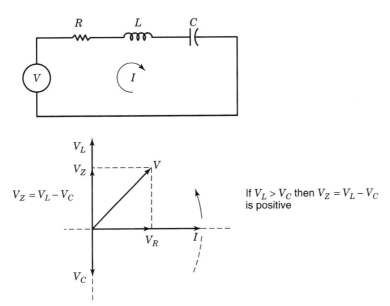

**FIGURE 3.3**   *RLC* series circuit.

is horizontal, as is the pointer for the voltage across the resistor. The pointer for the inductor voltage is vertical up and the pointer for the capacitor voltage is vertical down. If the reactance of the inductor is considered positive, the reactance of the capacitor is considered negative. These two pointers simply subtract from each other. In other words, the peak voltages across the capacitor and inductor are of opposite polarity at the same moments in time. The result of this subtraction can be positive or negative. If the inductive reactance dominate, the net reactance is positive. If the capacitive reactance dominates, the net reactance is negative. Above some frequency the inductive reactance dominates, and below some frequency the capacitive reactance dominates. The frequency where the two reactances cancel is called the *resonant frequency*.

The pointer system shown in Figure 3.4 can be used to determine the input voltage. The difference between the inductive reactance and the capacitive reactance is a vertical pointer. The voltage across the net reactance is also vertical. This pointer and the resistor voltage pointer make up the sides of a rectangle. The diagonal length is equal to the applied voltage. At resonance the reactances cancel and the diagonal length is the length of the resistor voltage pointer. In other words, the impedance of the series *RLC* circuit at resonance is equal to the resistance *R*.

At resonance there is current flow in both the capacitor and inductor. This means that there is energy stored in both components. The energy is transferred back and forth between the inductor and capacitor twice per cycle. This is exactly equivalent to a simple pendulum. When the pendulum is at the top of its swing, the mass stores potential energy. When the pendulum is at its

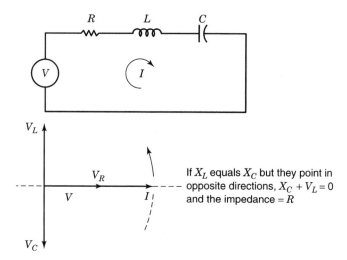

**FIGURE 3.4**   Pointer system applied to a series resonant circuit.

midpoint, the potential energy has been converted to energy of motion or kinetic energy. This transfer of energy occurs twice per cycle. The pendulum swings back and forth at its resonant or natural frequency.

The resonant frequency occurs when the reactance of the capacitor equals the reactance of the inductor. This is when $2\pi f L = 1/2\pi f C$. The frequency of resonance occurs at $f = 1/2\pi\sqrt{LC}$. The resonant frequency is in hertz when the inductance is expressed in henries and the capacitance is expressed in farads.

When an inductor and capacitor are placed in parallel, an applied sinusoidal voltage causes current to flow in both components. The reactances simply subtract and the resultant current for frequencies above resonance is capacitive and below resonance is inductive. At resonance the two currents are equal and subtract. This means that the voltage source supplies no current. In effect, the impedance becomes very large. If the inductance has a small series resistance, some current must be supplied per cycle. Assume that the reactances at resonance are 1000 Ω. If the impressed voltage is 10 V, the circulating current is 10 mA. If the series resistance is 10 Ω, the power loss is 1 mW. This power must be supplied from the 10-V driving source. This parallel resonant circuit looks like a resistance load of 100,000 Ω. This is indeed a high resistance. A series resonant circuit at resonance looks like a very low resistance.

## 3.5  PHASE

The pointer system that describes the voltages in an *RC*, *RL*, or *RLC* circuit illustrates that there is a time relationship between all the peak voltages and currents. The time relationships between sine waves in a circuit is called *phase*.

In a capacitor the voltage pointer is at right angles to the current pointer. This is called a 90° phase lag. In an inductor the voltage leads the current by 90°. The angles between the pointers are the phase shift in degrees. In the *RL* circuit there is a phase relationship between the driving voltage and the resultant current. This phase angle can be calculated directly from the geometry of the two pointers.

The current pointer was positioned horizontally in earlier discussions of the *RC*, *RL*, and *RLC* series circuits. This made it simple to determine the driving or input voltage. The inverse problem is usually specified. Given the input voltage the problem is to determine the amplitude and phase angle of the resulting current. The phase angle between the driving voltage and the resultant current does not depend on which pointer is horizontal. Any orientation can be used to determine the phase angle. The pointers can all be scaled so that the input voltage can be set to any given length. As an example, assume that the input voltage is 2 V when the current is 1 mA. The input voltage can be set to 5 V, and this scales the current level to 2.5 mA.

The discussion above assumes that the sine waves driving the circuits have been connected for a long time. When a sine-wave voltage source is first connected to a circuit, none of the components store any energy. After the circuit is operating for a time there is energy moving between elements in the circuit as well as energy supplied from the source. The pointer idea works only if the voltages and currents are sinusoidal and are the same in each cycle. There is a transitional (transient) period where the voltages and currents adjust to their steady-state condition. The rotating pointers, voltages, and currents all apply to this steady-state condition.

Impedance is a sinusoidal concept that applies to linear components. Engineers frequently apply the term to nonsinusiodal situations. It is a convenient way to express an idea, but it is not accurate. In a sense, impedance is a geometric concept, as voltages and currents need not be present to make the calculation.

## 3.6 PARALLEL *RL* AND *RC* CIRCUITS

The voltage impressed across parallel *RC* circuit causes currents to flow in both elements. The current in the resistor is in phase with the voltage, and the current in the capacitor leads the voltage by 90°. The two currents cannot add together arithmetically, as their two peaks do not occur at the same time. If the two current pointers are the sides of a rectangle, the diagonal length is the sum of the two currents. The angle between the diagonal and the voltage pointer is the phase angle of the resulting current. The reactance of the parallel circuit is the voltage divided by the length of the diagonal. The current in the parallel circuit leads the voltage by less than 90°.

Exactly the same thing happens for a parallel *RL* circuit. The current in the resistor follows the voltage, but the current in the inductor lags the voltage. If the two currents form the sides of a rectangle, the diagonal length represents the sum of the two currents. The impedance of the parallel circuit is the voltage divided by the length of the diagonal. The current in the parallel circuit lags the voltage by less than 90°.

## 3.7  PROBLEMS

1. A 10-kHz sine-wave voltage has a peak value of 10 V. What is the maximum positive slope of the voltage in volts per second? What is the minimum slope?

2. A 1-mH inductor is in series with a 0.01-$\mu$F capacitor. What is the resonant frequency?

3. If the inductor in problem 2 has a resistance of 1 $\Omega$, what is the reactance of the circuit at 90%, 100%, and 110% of the natural frequency?

4. A parallel resonant circuit consists of 1 $\mu$F and 1 mH. It has 10 V peak across the terminals at the resonant frequency. What is the current flow in the capacitor and inductor?

5. What is the phase angle between the current and voltage in a 10-$\mu$F capacitor at 1 kHz?

6. What is the phase angle between the current and voltage in a 1-H inductor at 100 Hz?

7. What is the impedance of a 1-$\mu$F capacitor in parallel with 1000 $\Omega$ at 1 kHz? At what frequency is the reactance of the capacitor equal to the resistor? What is the phase angle?

8. A 0.1-H inductor is in parallel with a 2000-$\Omega$ resistor. At what frequency is the reactance of the inductor equal to the resistor? What is the phase angle between the currents?

## 3.8  RMS VALUES

Voltage and current sine waves have been presented in terms of their peak values. A 10-V peak sine wave has a peak-to-peak value of 20 V. A 10-V peak sine wave voltage across a resistor will heat the resistor the same as a 7.07-V dc voltage. It is convenient to describe sine waves in terms of their equivalent heating value. Thus a 10-V peak sine wave is called a 7.07-V rms (root mean square) sine wave. A dc voltage of 7.07 V heats a resistor the same as a 10-V peak sine-wave voltage. The factor 0.707 is equal to $\sqrt{2}/2$. To obtain an rms value, the voltage or current wave is divided into small time segments. The squared amplitudes for each segment are summed, and the square root is taken of their average value. In most applications the letters *rms* are omitted and the sine-wave value is assumed to be rms.

A 10-V square wave has a peak-to-peak value of 20 V. The rms value for this square wave is 10 V. A waveform that goes from 10 V to 0 V 50% of the time heats a resistor 50% of the time. This waveform would dissipate 5 average watts in 10 Ω. A dc value of 7.07 V would also dissipate 5 W. This means that this square wave has an rms value of 7.07 V.

When two sine waves are added together at the same frequency, their phase relationship determines their rms sum. When the sine waves are at different frequencies, their heating ability is independent. Consider two sine-wave voltage sources in series heating a single resistor. The first one dissipates $V_1^2/R$, and the second one dissipates $V_2^2/R$. The total heat is $(V_1^2 + V_2^2)/R$. The rms equivalent voltage is $\sqrt{V_1^2 + V_2^2}$. This is how a true rms meter will measure two ac voltages in series at unrelated frequencies. This idea can be extended to any number of sine waves.

The concept of rms is used in many areas where heat is not involved. For example, a sinusoidal $H$ field is expressed in rms terms, not peak value. An $E$ field of 10 V/m at 1 MHz is understood to mean 10 V/m rms. The peak value is 14.14 V/m and the peak-to-peak value is 28.28 V/m. The rms idea is even used in treating the roughness of a surface. A surface is said to have an rms roughness of 0.1 cm. This means that the peak-to-peak surface variations are approximately 0.28 cm. The surface may vary in a random manner, but the rms concept can still apply. In this book, sinusoidal waveforms are given in terms of rms values unless otherwise specified. Two unrelated $H$ fields of 3 A/m rms and 4 A/m rms sum together to give 5 A/m rms. The expression rms would be understood and not printed.

## 3.9 PROBLEMS

1. A sine-wave voltage has a peak-to-peak value of 25 V. What is the rms value?

2. The power utility supplies 118 V rms for a business. What is the peak voltage?

3. A square-wave signal steps through the voltages 1, 2, 3 and 4 V. Each step lasts 1 ms. The steps continue to repeat. What is the rms value of this voltage?

4. Two unrelated sine-wave voltages sum to heat a resistor. The voltages are 6 and 8 V rms. What is the rms value of their sum?

5. A 12 V rms at 60 Hz voltage is used in a dc power supply. If the diodes used in bridge rectification require a forward voltage drop of 0.6 V, what is the resulting dc voltage? If the voltage sags 10% per half-cycle, what is the average dc voltage out? (*Hint*: A bridge rectifier uses two diodes in series in the forward direction for each half-cycle.)

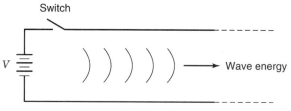

**FIGURE 3.5**   Simple transmission line.

## 3.10   TRANSMISSION LINES

The field energy that operates every component in a circuit must be carried between conductor pairs. It is important to understand how this energy is transported. Consider a battery, a switch, and a pair of parallel conductors, as shown in Figure 3.5. The moment the switch is closed, charges must appear on the surface of the two conductors near the switch. This flow of charge from the battery is a current. The presence of a charge on the conductors implies an $E$ field. The flow of charge establishes an $H$ field. These fields take time to appear, as energy cannot be moved in zero time. One way to understand the propagation of energy is to consider the inductance of the conductors and the capacitance between the conductors. In a very short unit of time, the current that flows charges a segment of capacitance and stores field energy in a segment of inductance. As time progresses, a second segment of inductance and capacitance stores energy. The energy for the third segment of inductance and capacitance must travel past the first and second segments. If the capacitances charge at a fixed rate, the current that flows is a constant. As time proceeds, electric and magnetic field energy is moving along the transmission line.

The field energy that fills the space between the conductors travels at about half the speed of light. In 2 ns ($10^{-9}$ sec) the field energy moves about 1 ft. If it were visible, a wavefront would be seen moving down the transmission line "pulling" energy behind it. The energy that is enroute cannot be lost. It must keep moving forward and go somewhere. It may be helpful to consider the head of the wave a railroad locomotive "pulling" cars.

In theory, this ideal wavefront would travel unimpeded for any distance. The source voltage supplies a steady current, which means that the transmission line looks like a resistor. If the transmission line is terminated in this resistance, the source will continue to supply a steady current as if the transmission line were infinite in length. This resistance is called the *characteristic impedance* of the transmission line. This impedance is a function of the geometry of the transmission line.

When the transmission line is left unterminated (open), the current at the end of the transmission line must be zero. The energy flowing in the transmission line cannot be stopped and this arriving energy must go somewhere. The forward wave is reflected so that the energy coming forward is directed back toward the source. This reflected wave cancels the current but doubles the voltage. Energy is now flowing in both directions simultaneously. When this reflected energy arrives back at the source, the energy again cannot be lost, so a second reflection must take place.

The second reflection is a wave that exactly equals the initial wave that started down the transmission line. The battery current is cut off, as the line cannot receive more energy. The voltage and current are now provided by the second reflection, not by the battery. This starts the process over again with wave energy that is going back and forth down the transmission line. The length of the energy train is equal to twice the length of the transmission line. This series of reflections is very ideal and assumes that there are no losses. In practice, the wave smears, there is radiation, there are losses, and the reflections die out. The end result is a fixed voltage along the entire line with no current flow. It is possible to see a few of these reflections using a high-speed oscilloscope.

When the transmission line is terminated in a short circuit, the energy going down the transmission line cannot be lost since $I^2R = 0$. This time a reflected wave is returned that cancels the voltage. When this wave arrives back at the source, a second reflection occurs. This time the battery must supply current for the initial wave, the first reflection, and also the second reflection. These reflections continue, with the battery supplying ever more current. In practice, the resistance of the conductors, the limitations of the battery, or a blown fuse terminates the current increase.

It is important to recognize that energy is transported between conductors in fields, not in the conductors. This means that the utility power we use comes to us via fields. Even the lowly flashlight battery supplies energy to the lamp via fields. This field view is not needed to analyze most circuits, but it is often the source of interference and noise. For this reason it is very important to keep this picture in mind. The issue of grounding becomes clear when fields are considered. A circuit view of grounding can be very misleading.

The textbooks treat a transmission line as a sequence of series inductors and shunt capacitors. The analysis divides the transmission line into a large number of very small components. This is called a *distributed parameter system*. This approach limits the fields to the components and does not allow for fields that extend out into the surrounding space. In practice, some of the field energy escapes the circuit and radiates. Any analysis that allows for this radiation is considerably more complex than the picture we have drawn.

## 3.11  POYNTING'S VECTOR

The energy that flows between two conductors requires both a magnetic and an electric field. The $H$ field is equated to the current flow and the $E$ field to the voltage across the lines. At all points in the space between conductors

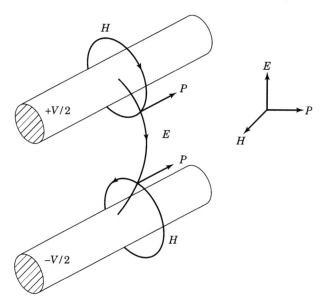

Note: $P$ is the power per unit area at a point in space.

**FIGURE 3.6**  Poynting's vector.

the $E$-field direction is perpendicular to the $H$-field direction. Both of these directions are perpendicular to the transmission path. The (vector) product of $E \times H$ at any point in space is called *Poynting's vector*. This vector points in the direction of energy flow. $E$ has units of volts per meter and $H$ has units of amperes per meter. The product is watts per meter squared at a point in space. Consider a plane that cuts perpendicular to the transmission line. This geometry is shown in Figure 3.6. Divide the plane into small regions where $E$ and $H$ are constant. Multiply the values of $E$ and $H$ times area in every region. Sum the results for every region over the entire plane. The result is the power supplied by the battery, which is volts times current. The power calculated using the fields agrees with the circuit approach. Point the two fingers and thumb of the right hand at right angles to each other. Let the first two fingers represent the $E$- and $H$-field direction, respectively. The thumb will point in the direction of energy flow. If the $E$ field is reversed in direction, the hand must turn over to represent the new field. Now the thumb points in the opposite direction. This means that the direction of energy flow is reversed.

## 3.12   TRANSMISSION LINE OVER AN EQUIPOTENTIAL SURFACE

The $E$- and $H$-field pattern in Figure 3.6 is symmetrical about a plane that is midway between the two conductors; this is shown in Figure 3.7. The

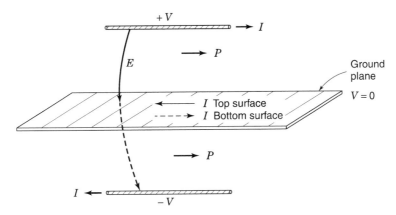

**FIGURE 3.7** Transmission line over an equipotential surface.

plane is an equipotential surface that is at 0 V. The $E$ field is everywhere perpendicular to this surface. If this equipotential surface is replaced by a conductive surface, the energy is transported in the forward direction above and below this surface. Note that there is equal current flowing in opposite directions on the two sides of this added conductive surface. If the bottom transmission line conductor is removed and the added conductive surface is connected to 0 V, the top half of the transmission line will carry exactly one-half the energy. Now current flows only on the top of this added surface. The current flow is concentrated where the terminating $E$ field lines are closest together. For a transmission line terminated in its characteristic impedance, the current that flows is constant, yet the current pattern is established by the $E$ and $H$ fields. If the transmission line is brought closer to the equipotential surface, the intense portion of the field is confined to a smaller volume and the current flow is restricted to a narrower path. This is an important fact that is not considered in circuit theory. Since the current does not use the entire equipotential surface, the effective resistance is higher than the available resistance. The characteristic impedance of the transmission line in Figure 3.7 is one-half the impedance of the full transmission line of Figure 3.6. This results directly from the fact that the voltage is one-half and the current is unchanged. The equipotential surface is often called a *ground plane* or a *ground reference plane*. In many cases the ground plane is not flat, but the term *plane* is still used.

## 3.13 TRANSMISSION LINES AND SINE WAVES

Transmission lines are often used to carry power using high-frequency sine waves. An example might be a line from an AM radio transmitter to an antenna. The problem of power transfer requires impedance matching. When the voltages and currents are sinusoidal, the input impedance of a transmission line depends on frequency, line length, line termination, and characteristic impedance (geometry). Transmission lines can only transport field energy; they cannot dissipate energy. In an ideal line, energy can only be dissipated in a terminating resistor.

Pairs of conductors abound in all electrical processes. Conductor pairs include the earth, ground planes, shields, power conductors, electrical conduit, building steel, racks, gas lines, telephone lines, and so on. The list extends to all metal structures. All conductor pairs can support the movement of field energy. In general, it takes less energy for a signal to follow the space between conductors than to travel across free space. This means that the path for interfering signals is modified by the presence of conductor pairs, and unwanted energy can be brought into circuits that may need protection. These signals may be sinusoidal in character or they can be described in terms of sinusoids. For this reason it is important to understand the character of ideal transmission lines where sine waves are involved.

The reflections that take place at the ends of a transmission line occur for all waveforms. At an open circuit the reflected wave must cancel the current. At a short circuit the reflected wave must cancel the voltage. Any reflected sine wave returns back to the source carrying energy with it. The source is also a reflection point where the voltage is defined. If the returned sinusoidal signal is equal to the driving signal, the driving current that flows will be zero. In other words, the transmission line looks like an open circuit. If the returned sinusoidal signal is at 0 V when the driving signal is at its maximum, the transmission line looks like a short circuit. If the voltage that returns back to the source is at some other phase angle, the driving voltage supplies a reactive current to the line. For lines terminated in open or short circuits, the input impedance of the line can only appear to be a reactance, as there can be no energy lost in the line. Energy cannot be lost in an open or short circuit. The input impedance must therefore be a capacitive or inductive reactance, depending on the length of the line. If energy is dissipated at the terminating end of the line, the reflected wave will change in character. The current supplied to the line will adjust in phase to supply this energy. The input impedance is the reactance of the unterminated line in parallel with the terminating load resistance.

Consider an open transmission line at 1 MHz. Assume that the signal travels at a speed of 150 m/$\mu$s. If the length of the line is 75 m, the reflected signal arrives back in one cycle. This means that the input voltage equals the returning voltage and the input impedance is high. If the line length is 37.5 m, the input impedance looks like a short circuit. For line lengths below 37.5 m, the line looks like a capacitive reactance. If the line is terminated in a short circuit, the line looks like an inductive reactance for line lengths below 37.5 m.

Unintentional transmission lines rarely have nice geometries. This simply means that there are many complex reflections. These conductor geometries support the flow of field energy, and this energy, once captured, reflects back and forth until lost in the parasitics of the system. In many situations a balance is reached between arriving energy and dissipated energy. Parasitics include conductor losses and reradiation. This is no different than the absorptions and reflections that occur when a light illuminates a room. Every surface absorbs and reflects light energy, resulting in an ambient light level. The resultant light intensity at any one point is not easily calculated. In a typical facility there are ambient electromagnetic fields that are the result of wave propagation and reflections from the many conductor pairs.

The transport of all electrical energy or information is via some sort of field. The field is often associated with a transmission line. In most applications where the frequencies are below 100 kHz, a circuit theory approach is the most practical way to analyze a process. The reflections when they occur last a small fraction of a cycle and can be ignored. The dimensions of a power grid can be hundreds of miles, and here the reflection processes must be considered.

## 3.14   COAXIAL TRANSMISSION

The transmission lines discussed in Section 3.13 consisted of parallel conductors or conductors over a ground or earth plane. Another geometry that is frequently used is the *coaxial cable*. This cable is similar to the single-conductor shielded wire discussed in Section 2.9. In a high-quality coaxial cable, the geometry of the cable is held constant so that the characteristic impedance is controlled. The dielectric is carefully selected and the center conductor is kept concentric with the shield. In high-quality coax the inner conducting surface is almost mirrorlike in character.

The characteristic impedance of coaxial cables varies from 30 to about 75 $\Omega$. Parallel open conductors have characteristic impedances that are generally higher. For example, television antennas lead-in wire is rated at 300 $\Omega$. The is a good impedance match for the antenna and for the input circuits in the television set. This is a case of transferring a maximum amount of energy from the antenna to the set.

Coaxial transmission has the advantage of fully containing the electric and magnetic fields of transmission as shown in Figure 3.8. The $E$ field terminates on the inside surfaces of the cable. The current flows out on the center conductor and returns on the inside surface of the shield. For this current pattern, Ampère's law applied around the entire coaxial cable requires that the $H$ field be zero. This means that the energy that is transported inside the cable generates little if any magnetic field outside the cable. An external magnetic field (induction field) can still enter the space between conductors inside the coaxial cable.

The essential difference between a coaxial cable and a shielded cable is in application. The shield must be a return path for current flow in a coaxial

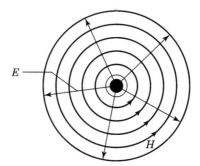

**FIGURE 3.8**   *E* and *H* fields in a coaxial cable.

transmission. In applications requiring shielding, the outer conductor may be connected to a reference potential at only one point. The location of this one point is critical and is discussed later. Shielded cable is best used in applications below 100 kHz. Coaxial cable can be used to handle signals that extend up to several gigahertz (1 gigahertz is 1000 MHz). For short distances (a few feet) the type of cable that is used is usually not critical. The termination of coaxial cable can be troublesome, however. It is in this area where field leakages or field couplings are most apt to occur. This topic is discussed in later chapters.

## 3.15   UTILITY POWER DISTRIBUTION

Utility power is supplied over transmission lines from generators. A typical utility generator develops sinusoidal voltages on three conductors with earth as the reference conductor. The voltages reach their positive peak values separated by 120 electrical degrees. At 60 Hz one full cycle takes 16.7 ms. This means that one of the power conductors reaches a positive peak every 5.56 ms. This three-phase power generation is very effective in that the energy flowing in the field is constant at every moment in time. This means that the torque on a generator shaft does not vary over any one cycle.

Insulator breakdown limits the maximum voltages permitted in generator design. The magnetic field between conductors tends to push the conductors apart. During a fault condition, high current flow can destroy the generator. For this reason fault protection is a key factor in generator design. The neutral is often grounded through a low-ohmage power resistor. If an ungrounded power conductor should fault to ground, the current is limited by this neutral grounding resistor. This provides mechanical protection. The idea is to remove the load before there is overheating.

Transformers near the generators step up the voltages so that the currents in the transmission lines are at acceptable levels. At local distribution points, transformers step the voltages back down to lower levels. In an industrial

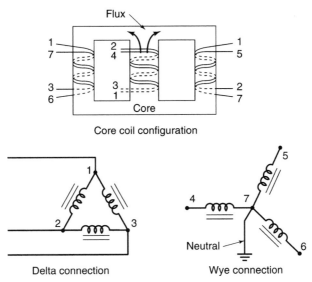

**FIGURE 3.9**   Delta–wye transformer circuit.

area, site transformers are installed that supply various voltages for motor loads, lighting, and general use. In residential areas, power is brought into residences at a nominal 120 V (rms). Often, the power is supplied from a transformer winding that is centertap grounded (earthed). This arrangement provides two sources of 120 V. The voltage across the entire transformer leg is 240 V. This higher voltage can be used to power an electric range, for example.

Three-phase power distribution transformers are built on magnetic cores with three magnetic paths. Each path has a coil of the primary and a coil of the secondary. The primary coils can be connected in a ring called a *delta connection*. The three secondary coils are usually connected in a *wye configuration*. These coil arrangements are shown in Figure 3.9. Power is taken from the secondary of the last transformer from connections made across each coil. The common or midpoint of the wye is called the *neutral*. If this neutral point is brought out as one of the power conductors, it is called the *neutral conductor*. The loads across each coil share this one common connection. If the loads are all equal and demand sinusoidal current, the total neutral current will sum to zero.

Lightning protection for a transmission line is a very important consideration. The utilities do not want to lose transformers or switching gear along the transmission path. Protection is provided by an earth connection to the neutral at every transformer and power pole. This provides a lightning path to earth outside the transformer. For power entrances into facilities, the neutral conductor is earthed at each service entrance. A lightning pulse that strikes the power conductors has a convenient path to earth outside the facility. If a lightning pulse were to enter a transformer or a facility, it could cause damage and be very dangerous.

The neutral in a typical distribution scheme is multiply earthed. Loads are rarely balanced and this means that there is some net neutral current flow. This current flows in both the earth and the neutral conductors provided. Even if the earth current is a small percentage of the total current, it can still be a source of interference. The earth is a relatively high resistance and neutral currents will often follow nearby conductors that happen to be buried in the earth. This might be building steel, conduits carrying water or gas, or even steel fence lines. This means that power currents and their associated fields are a part of the ambient in any facility. It is difficult to predict the patterns of these current loops and to calculate their field strength. The 60-Hz field is usually not the issue. The neutral currents that flow are often rich in harmonics, which means that the current has content that is changing quite rapidly. This means that the resulting magnetic fields are more easily coupled into cables and wiring associated with various pieces of electronic hardware.

Any conductive loop is susceptible to coupling from fields. It is obvious that power distribution bring more than power into a facility. Remember that power is not carried in conductors. The loops that involve neutral current in the earth are apt to be physically large. The earth becomes one of the conductors involved in energy transfer. If this same space contains other interfering fields, these signals will also couple to the power conductors, which in turn connect to every piece of hardware.

An interference field in a facility causes current flow in all conductors. This current is easily detected by placing a current probe around a facility conductor. From a circuit analysis viewpoint the voltages producing this current are not locatable. A changing field is the only evidence of an interfering source In fact, eliminating this field source is usually impossible. If the interference is troublesome, some change in geometry is required so that the field is not coupled into critical circuits.

## 3.16  EARTH AS A CONDUCTOR

Earth resistances are relatively high compared to a metal conductor. Consider two conductors that are buried deep in the earth spaced apart by 100 m. The resistance between the two conductors is twice the resistance of the one earth connection. Under ideal conditions with a chemically treated well, a resistance below 1 $\Omega$ is difficult to achieve. In neutral grounding for a residence, a resistance of 20 $\Omega$ is all that is required to meet the safety code. It is important to remember that this is a low-frequency impedance.

It is illegal to have two separate and unconnected earth grounds in one facility. Consider a fault condition where a "hot" conductor contacts a second earth. At 120 V, an earth resistance of 10 Ω represents a load of 12 A. This low a current will not trip a breaker. After the fault occurs, conductive materials associated with this earth will be 120 V from conductive materials contacting the first earth. This represents a severe shock hazard. The National Electrical Code (supported by the National Fire Protection Agency) requires all building conductors that could contact any electrical wiring be a part of one equipment grounding system. This requires intentional bonding between various facility conductors. To trip breakers, the fault impedance needs to be milliohms. The equipment grounding system can make many earth connections as long as the conductors are all bonded together. The neutral or grounded conductor carrying power can only be earthed once and this is at the service entrance. If the neutral is multiply earthed in a facility, the fault protection system may not function. The equipment grounding system includes all conduit, visible building steel, motor housing, metal siding, and so on. These grounds (conductors) must be bonded together and bonded to the neutral at the service entrance.

## 3.17  POWER TRANSFORMERS IN ELECTRONIC HARDWARE

The power utility supplies the energy needed to operate most electronic circuits. This utility power is brought inside racks on power strips or to wall receptacles for convenient connections to hardware. Usually, three conductors are involved: the ungrounded conductor (*hot lead* or 120 V), the grounded conductor (120 V return), and an equipment grounding conductor (*green wire*). The primary coil of a power transformer is connected between the 120 V and the grounded conductor. The equipment ground conductor or green wire is a safety conductor that connects to the metal cabinet if it exists. The transformers inside the hardware have secondary coils that provide voltages for the various power supplies used by the circuit.

The secondary coils provide ac voltages that can be rectified and filtered. These rectified voltages (dc supplies) are floating, which means that as a circuit they can be referenced to a second ground potential. Diodes connected to the secondary coils provide rectification, which permits current to flow in one direction. This current flows into filtering electrolytic capacitors, which store field energy over the power cycle. The circuit can draw energy from the capacitor over the entire power cycle. A typical rectifier system generating a plus and minus dc supply is shown in Figure 3.10. Rectifier systems should not draw an average dc current from any coil. This can saturate the transformer core material, resulting in excessive magnetizing current.

In most designs voltage waveforms across filter capacitors peaks twice per cycle. The capacitors are recharged whenever the coil voltage exceeds the capacitor voltage. This peaking of voltage across the capacitor is called *power supply ripple*. The average sag in the voltage is the result of the circuit drawing

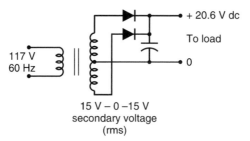

**FIGURE 3.10**   Typical rectifier system on the secondary coil of a power transformer.

current from the capacitor. The peak dc voltage is the peak ac voltage minus the voltage drops across series diodes. For silicon diodes this is about 0.6 V per diode.

The current that flows in the primary of the transformer must follow the pulse character of the secondary current and peak twice per cycle. The only time that energy is supplied from the utility is in these peak periods. These pulses of current imply a pulsed magnetic field around the power conductors. This type of rectification can exist in many pieces of hardware, with the result that the peak current in the power conductors can be quite high. When these peaks of current flow in the neutral conductor, there is a resultant neutral voltage drop. The neutral conductor is no longer at the earth potential or at the equipment ground potential. In three-phase systems these pulses of current do not cancel each other as the pulse timing is wrong. This means that the neutral current can be quite high, as it results from nonlinear loads on all three phases. High neutral current has been a source of difficulty in many electronic installations.* One solution is to use a local separately derived three-phase transformer. This transformer does not require a neutral connection to its primary. The neutral run on the secondary can be kept short, eliminating a large neutral voltage drop. If necessary, several separately derived transformers can be used in one facility. Neutral grounding on the secondary is required and must meet code requirements.

Each transformer in a hardware complex has a reactive path between the primary and secondary coils. This capacitance on the average will be about 250 pF coil to coil. This is a reactance of 10.6 MΩ at 60 Hz. Transformers are built with one end of the primary coil next to one end of the secondary coil. If the ungrounded conductor is the end next to the secondary coil and the secondary circuit common is remotely grounded, it is possible for the primary voltage of 120 V to cause 1.3 mA to flow in this connection. If the

---

*High neutral current is a problem, as this conductor is one side of the utility power brought to each piece of hardware. Most hardware transformers are unshielded and this allows the neutral voltage to cause interference currents to flow through leakage capacitance across the windings to secondary circuits.

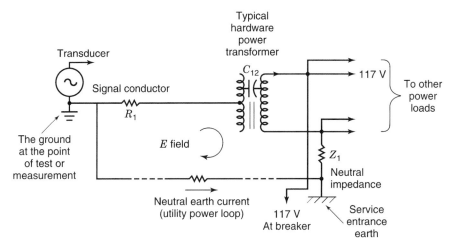

**FIGURE 3.11**   Loop involving grounds and a power transformer.

primary voltage is 480 V, the current is obviously higher. Figure 3.11 shows the voltages in a loop involving the earth ground, the circuit ground, and the line voltage. Note that there is no one point in a facility that could be properly called the zero of potential. The loop area in Figure 3.11 is a function of the wiring in the facility. This loop also couples to any fields that cross this area. If this loop is associated with a signal of interest, the voltage drop on part of the loop will interfere with this signal of interest. If the resistance of the conductor is 10 $\Omega$, then a current flow of 1.3 mA will introduce a signal of 13 mV. For a rapidly changing field, this voltage could easily exceed 2 V, which is enough to damage some unprotected circuits. In many analog systems, 10 mV is a full-scale signal. The peak noise signal in this situation might be specified at 10 $\mu$V.

## 3.18   ELECTROSTATIC SHIELDS IN ELECTRONIC HARDWARE

The idea of using a shield for electrostatic interference was introduced in Section 2.9. Placing a circuit entirely inside a conductive box limits the effects of external electric fields. In the case of a metal box, complete shielding is violated at the transformer. At first glance it would seen appropriate to place a metal shield between the coils of the transformer and connect it to the box. This would place the primary coil and its voltages on the outside of the box and the secondary coils and their voltages on the inside of the box. These shield arrangements are shown in Figure 3.12. There are several problems associated with this approach. If the shield is connected to circuit common inside the box as in Figure 3.12a, the primary coil can still circulate current in the secondary circuit common. If the shield is connected to the box as

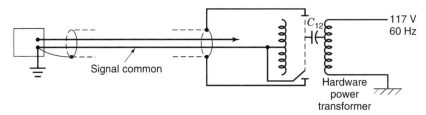

(a) Primary current circulates in signal common through $C_{12}$

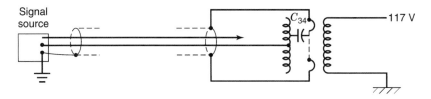

(b) Secondary current circulates in signal common through $C_{34}$

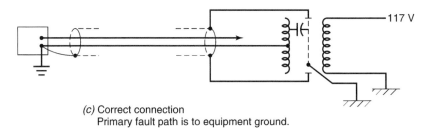

(c) Correct connection
Primary fault path is to equipment ground.

Note: Transformer core symbols are omitted for clarity

**FIGURE 3.12**    Single shield added to a hardware transformer.

in Figure 3.12b, the secondary coils will circulate current in the secondary common. A serious problem occurs if there is a power fault condition in the transformer. The fault path is from the primary coil through a signal ground into another piece of hardware. The only way to avoid this condition is to connect the shield to the green wire or equipment-grounding conductor. Now the fault current flows in a path designed to trip a breaker. For this reason, single shields in power transformers for electronic hardware should always be connected to equipment ground (green wire) as in Figure 3.12c.

Shields in transformers are often called *Faraday shields*. The shield material can be a single layer of copper or aluminum. Care must be taken so that the shield does not form a shorted turn. If the shield material folds down over the sides of a coil, it is called a *box shield*, which further limits the leakage capacitance. Coil-to-coil capacitances can be around 500 pF. With a simple shield the leakage capacitance might be 5 pF. With a box shield the leakage might be below 0.2 pF. In this last arrangement the leads themselves might have to be included in the shield.

## 3.19   WHERE TO CONNECT THE METAL BOX

Experimenters frequently place circuits inside a metal box. These circuits are often analog in nature where there is a need for signal amplification. The unenclosed circuit picks up "noise and hum" from the environment, and the metal box acts as a shield reducing this interference. This approach seems to work best when the box is connected to the reference conductor in the circuit. When the box is left floating (not electrically connected), it assumes some average room potential that is not the same as the potential of the reference conductor of the circuit. This allows current to flow in mutual capacitances that couple interference into the circuit. There is a mutual capacitance from the box to the input of the circuit and a mutual capacitance from the output of the circuit to the box. This amounts to feedback, which can modify circuit performance and cause instability. The experimenter finds that many problems disappear when the box is connected to the circuit common or reference conductor. Often, little attention is paid as to where the connection is made. The circuit and some of the key mutual capacitances are shown in Figure 3.13*a*. In Figure 3.13*b* the box is grounded (connected to the circuit common or reference potential), thus changing the nature of the mutual capacitances. The feedback path is no longer a threat and the metal box does not couple undesirable "room pickup" currents into the mutual capacitances.

Assume that an input signal line is associated with this battery-operated circuit. One side of the signal line is grounded. This means that the input signal common connects to a second circuit common at some remote point. This remote point may or may not be earthed. The metal box must extend over the input leads, or the electric fields in the area can cause current to flow in these leads. The input lead that is not grounded is usually more sensitive to pickup than is the common lead. The signal of interest is the potential difference between these two input leads. The extension of the metal box is usually a braided shield over the two signal conductors. This circuit is shown in Figure 3.14.

There are two choices as to where the metal box and shield connect to the circuit common. When the connection is made inside the box at the circuit common, there is a path for the interference current to flow in the grounded reference conductor. This path is numbered ①⑤③④① as shown in Figure 3.15*a*. To avoid this problem, the shield (the extension of the box) must be connected to the signal common at the end of the signal cable as in Figure 3.15*b*. Now external fields cause currents to flow in the numbered path ①⑤④① and this excludes the signal conductor. This little exercise yields two rules:

1. A metal enclosure that shields a circuit should be connected to circuit common.

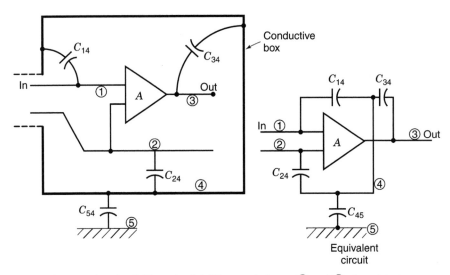

Note: A is any active circuit. The potential difference between ④ and ⑤ allows interference
    currents to flow in the circuit

*(a)*

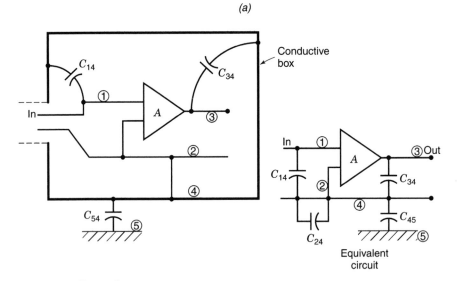

Note: When ④ and ⑤ are connected $C_{24}$ is shorted out. Feedback path $C_{14} \rightarrow C_{34}$ is
    eliminated. Current flowing in $C_{45}$ does not flow in the parasitic capacitances of the
    circuit.

*(b)*

**FIGURE 3.13**   Mutual capacitances and the metal box.

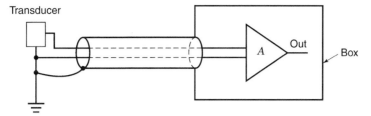

**FIGURE 3.14**   Grounded input signal brought into a circuit in a metal box.

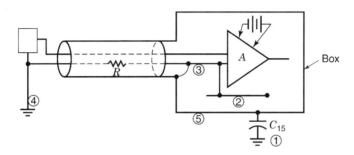

*(a)* Incorrect; circulation is through $R$. ①⑤③④①

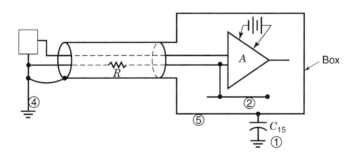

*(b)* Correct; circulation is ①⑤④①

**FIGURE 3.15**   Grounding of the box.

2. The connection should be made where the common conductor connects to an external ground.

These rules are not obvious from looking at a circuit diagram. The correct connection is a matter of geometry. The question is not whether to make a connection but where to make the connection.

In many circuit applications the shield or box can be connected without regard to these rules. This simply means that the signal levels are high enough

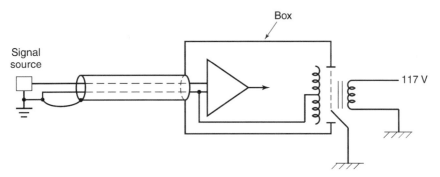

**FIGURE 3.16**   Grounded input lead and a power transformer in a metal box.

so that the interference that results is of little importance. It is always wise to look at the interference level that might result if the rules are not followed. The best method is to actually measure the performance both ways.

The circuit in Figure 3.15 is battery operated. If an unshielded power transformer is placed back into the circuit, power current will circulate from the primary coil to the circuit common ground. This current flows in the signal common lead and can be a source of interference. The interference is reduced if the grounded side of the power source is next to the secondary coil. The interference is reduced further if there is a transformer shield as shown in Figure 3.16. Typically, a shield will limit the current flow because the mutual capacitance from the primary coil across the shield is about 5 pF. Also note that the single shield is not at the same potential as the circuit common, and this potential difference can also circulate interference current in the input signal common.

This interference problem is not a simple one. A single grounded connection to a circuit powered from a shielded transformer will carry interference current. The interfering sources are:

1. Neutral voltage drop
2. Potentials between the equipment grounding conductors and the neutral conductor
3. The coil potentials in the transformer
4. Leakage capacitances through and between shields

Thanks to transistors and integrated circuits, there are many circuit techniques that sidestep this entire problem. For a short signal run the interference is minimal. In most digital applications the interference level is far smaller than the signals of interest; thus there are few problems. The author has seen cases where signals were connected between two large computers. Transients in the facility were coupled into this signal loop via the power system, which resulted in damage to signal circuits (see Section 7.4).

In analog work, noise coupling is a serious problem. In the early days of vacuum-tube design, one solution that seemed to work was to connect the metal box to the equipment ground. A water pipe ground was brought out at every laboratory bench. A connection to this ground placed the transformer shield at the same potential as the metal box, thus eliminating one source of interference. This was fine as long as the signal ground was at or near this same potential. This solution has many drawbacks and is not recommended.

Grounding equipment housings to equipment ground or better yet, finding "good grounds," has been a quest for many engineers. The idea is simple enough. The equipment grounding conductor is earthed. If an earth ground connection reduces the system noise level, a better earth ground should reduce it further. This leads to the idea of making an excellent contact with the earth and grounding the entire facility (electronically) to this one point.

The trouble with this viewpoint is that it ignores the problem of limiting interfering current in sensitive leads. This approach does not control the electromagnetic fields that can couple into sensitive areas. There is nothing wrong with a good earth connection, but the question still remains as to how to build electronics that is consistently free from interference. The "good ground" is expensive and is seldom used to any advantage. This one-point ground cannot be used effectively to protect a facility against a lightning strike. Lightning protection still requires many down conductors to carry lightning current to earth. Nature will probably not use this one central (easy) path to earth. The sign reading "Path for Electronic Interference" will probably be ignored. A good ground (earth) can in no way eliminate electric or magnetic fields in the area.

## 3.20 PROBLEMS

1. A 10-V voltage pulse lasts 0.001 s and recurs every 0.01 s. What is the rms heating?

2. A transmission line is 10 m long. How much time does it take a reflected pulse to return to the source?

3. A transmission line has a characteristic impedance of 50 $\Omega$ and the line is shorted at the end. The line is 10 m long. How much current is flowing after 0.3 $\mu$s if 10 V is switched on the line?

4. A steel structural member in a building carries 2 A at the fifth harmonic of the power frequency (300 Hz). What is the $H$ field at a distance of 1 m? At this distance, what is the voltage induced in a loop having an area of 0.1 m$^2$?

5. A 10-$\mu$F filter capacitor is charged from a 10-V ac source through a silicon diode. What is the peak voltage? If the capacitor is discharged at a steady rate of 10 mA, what is the voltage after 10 ms? Assume that the capacitor is not being charged after the peak voltage occurs.

**6.** A dc source of 2 V and a series ac source of 3 V heats a resistor of 10 Ω. What is the power dissipated?

**7.** Two sine-wave voltages of 10 V peak are summed. Their phase relationship is 90°. What is the total voltage? What is the total voltage if the phase angle is 45°?

**8.** A transformer shield has a leakage capacitance of 5 pF. A pulse on the power line is rising at 10 V/μs. What is the current level of the pulse on a grounded secondary conductor? What does the current pulse look like if the voltage pulse rises for 5 μs and falls for 5 μs?

## 3.21  REVIEW

Sine waves are at the heart of describing electronic behavior. When a sine wave in impressed on a linear circuit, all the currents and voltages will be sine waves. Capacitors and inductors offer opposition to sine-current flow. This opposition is called reactance. Combinations of resistance and reactance are called impedances. A simple geometry can be used to see how voltages and currents add together when they involve reactances and resistances. Resonance occurs in circuits when the inductive reactance equals the capacitive reactance.

The transport of energy between conductors involves electric and magnetic fields. The power density at a point is given by Poynting's vector. Pairs of conductors abound in a facility carrying field energy into hardware. Coaxial cable can be used to confine field energy so that it does not radiate or reflect along its path. Power transformers used in hardware are a connection to the power grid. This connection generates currents that can flow in signal, input, output, and control conductors and is a source of interference. Safety dictates that a single shield in a power transformer be connected to equipment ground. The box or enclosure for sensitive electronics should be grounded where the signal grounds. Any other connection point allows the fields in the area to circulate currents in sensitive leads.

# 4 A Few More Tools

## 4.1 INTRODUCTION

The behavior of circuits at frequencies above 1 MHz is complicated by radiation, reflections, and skin effect. At higher frequencies, circuit theory predictions can become quite inaccurate. Circuit symbols become less and less respresentative of circuit behavior and circuit geometry becomes the entire issue.

Geometry plays a role in all design. In low-level analog instrumentation the secrets of good performance are in component and trace location. A schematic may not provide sufficient information for the product to be duplicated correctly. Many aspects of design evolve as a product matures. Design difficulties can be avoided if a basic understanding is in place. In early radio design, circuits were separated into metal compartments to avoid unwanted feedback. Compartments were expensive and bulky. Eventually, better circuit techniques allowed these compartments to be discarded. Products often have clumsy beginings. Often, design weaknesses are not corrected until late in a product's evolution.

It is difficult to define the electrical performance of a facility when it involves the control of noise and interference. The electrical performance of a building design is rarely considered, although precautions are often taken. Many precautionary design details are unnecessary, but it is very difficult to argue pros and cons without some basis of understanding.

Electronic design often involves the electrical behavior of conductors and insulators that are not components. A ground plane, the earth, thin conducting sheets, and long conductors are examples of such items. The following sections discuss tools that are needed to handle these conductor geometries.

## 4.2 RESISTIVITY

The resistance and impedance of various conductors plays a role in electronic design. At low frequencies, a conductor is said to have a resistivity given in units of ohm-centimeters. A simple example can illustrate how this resistivity measure is used. Copper has a resistivity $\rho$ of 1.72 $\mu\Omega$-cm. The resistance of a round wire can be obtained by multiplying the resistivity by the wire length and dividing by its cross-sectional area. A copper wire 1000 cm long with a

**75**

cross-sectional area of 0.01 cm has a resistance of 0.176 Ω. This resistance value assumes that the current under consideration flows uniformly in the entire wire. A good benchmark to remember is that 1000 ft of No. 10 AWG (American Wire Gage) copper has a resistance of 1 Ω. The resistance doubles for every increase in three wire sizes.

It is interesting to consider the resistance of earth connections. The resistivity of earth varies depending on soil type and moisture content. For damp claylike soil the resistivity might be 10,000 Ω-cm. A cube of soil one meter on a side has a resistance of 100 Ω. The resistance of a cube of soil 10 m on a side is only 10 Ω. This measure assumes that a uniform ideal contact exists over two opposite faces of the cube. In this second case the contact area must be 200 m². This fully illustrates the problem of making a low-resistance earth connection. A chemical well is one way to obtain a large contact area. The well is filled with a conductive salt that provides a large contact area with the surrounding earth. A center conductor connects to the salt. Any human-generated current flowing into the earth must exit the earth somewhere. The resistance of a full circuit thus requires two such high-quality contacts.

The resistivity of desert sand, solid granite, or of a lava bed is very high. It is not practical to seek an earth ground on these surfaces. Power safety and lightning protection requires that all conductive elements in a facility be bonded together to form one equipment grounding conductor. The neutral power conductor must be connected to this conductor at one point, the service entrance. If handled correctly, a facility built on an earth insulator can be made perfectly safe. Don't forget that the electronics on board an aircraft works fine without an earth connection. The fuselage is floating from earth and represents the local ground plane. Aircraft can receive direct lightning hits and still not malfunction.

## 4.3 INDUCTANCE OF ISOLATED CONDUCTORS

A conductor carrying current implies a magnetic field. The flux generated per unit of current and per unit of length represents an inductance per unit length. This magnetic flux is essentially independent of conductor diameter. The equations for calculating this inductance are complicated by the fact that all circuit currents must flow in some sort of a loop. Assigning an inductance to an isolated section of conductor requires some assumptions. For this book it will suffice to provide a starting point for estimating the inductance of typical isolated conductor runs. Ten inches of 5-mil-diameter wire (No. 36) has an inductance of 0.4 μH (a mil is 1/1000 of an inch). Ten inches of 257-mil-diameter wire (No. 2) has an inductance of 0.2 μH. The inductance is essentially proportional to wire length. A 50-fold increase in wire diameter only halves the inductance. The only practical way to lower this inductance is to provide parallel conductors.

## 4.4 OHMS PER SQUARE

The resistance of a square of conductive material is equal to the resistivity of the material times the length of the square divided by the area of one edge. This assumes that current flows uniformly across the square from edge to edge, not surface to surface. The resistance across the square is the resistivity of the material divided by the thickness of the square. Notice that this resistance is independent of the dimension of the square. This means that any square of the same material and thickness has the same resistance. Again this assumes that the current flows uniformly across the square from edge to edge. The resistance of every square of copper 0.01 cm thick is 172 $\mu\Omega$. This is a very low resistance considering that this could be a square the size of a postage stamp or of a large building. The resistance of every square is the resistivity $\rho$ divided by the thickness in centimeters.

The resistance of a rectangle can be calculated by dividing the area into squares. A 10 ft × 30 ft rectangle of copper 0.01 cm thick has a long dimension resistance of 516 $\mu\Omega$. In this case there are three squares in series. The short dimension resistance is 57.3 $\mu\Omega$. Here the three squares are in parallel. In both cases the current must flow uniformly across the rectangle.

## 4.5 PROBLEMS

1. What is the dc resistance of a copper wire 1 m long with a diameter of 0.1 cm?

2. A soil sample has a resitivity of 25,000 $\Omega$-cm. What is the resistance across the faces of a cube 5 m on a side?

3. Estimate the inductance of an isolated 5-mil copper wire 10 ft long.

4. What is the resistance of a rectangular sheet of copper 10 m long, 1 m wide, and 0.02 cm thick?

5. What is the ohms-per-square resistance of a square of steel 0.01 cm thick with a resistivity of 10 $\mu\Omega$-cm?

## 4.6 RADIATION

The field patterns that are shown for various conductor configurations extend out into space. In the case of a coaxial cable geometry, the fields are confined inside the cable. In the circuit view of electronics, field energy for inductors and capacitors does not leave the confines of the components. In the real world as the components get smaller in value, a larger percentage of the field energy storage is in the space around the components. The conductors that interconnect circuit elements also have fields that extend out into space.

The statement that the fields extend out into space must be explained. Fields in space represents energy and energy cannot be moved in zero time. At the speed of light it takes about 1 ns for an electromagnetic field to move 1 ft. The circuit concept for an inductor or a capacitor requires that all resulting field energy be returned to the circuit twice per cycle at the sinusoidal frequency of interest.* What happens if the frequency of interest is so high that the energy cannot return in phase with the current or voltage generating the field? This energy is said to be *radiated.*

When field energy leaves a circuit, what tells it to return to the circuit? One explanation that might be used is the idea of *reflection.* Divide space into a cluster of cones that emanate from the area generating the field. As the diameter of the cone increases, the characteristic impedance of this transmission space changes. The outgoing wave reflects a small amount of energy at every surface that crosses the cone. In effect, there is field energy moving forward and backward for every point in space. This complex motion of field energy can be predicted from fundamental field relationships. These relationships are known collectively as *Maxwell's equations.* It is interesting to note that all we know about electrical behavior was explained by these equations, which were written in the mid-nineteenth century.

Radiators in daily operation carry data, voice, and video. The lowest-frequency radiators are around 10 kHz and are used to communicate to underwater craft. Radar transmitters work in the gigahertz. The majority of commercial transmitters are above 500 kHz. The most powerful radiators are military radar transmitters operating in the gigahertz. In an urban environment, all areas are constantly being bombarded with these electromagnetic radiations. Intense radiation can be used in an attempt to disrupt military aircraft. Understanding radiation and understanding radiation hardening (shielding) is an important topic.

## 4.7  HALF-DIPOLE ANTENNAS

Developing a theory of antenna radiation requires many simplifications. Designing radiating antennas is not the aim of this book. Our interests relate to the field strengths of existing transmitters (interference), the fields that are inadvertently radiated by circuits and the impact of radiation that enters a circuit. The intent in this book is to avoid building electromagnetic radiators.

The analysis of a dipole antenna starts out by dividing the antenna into individual current elements. These elements of current cannot exist any more than the element of current that was used to explain magnetics. This dipole element has zero resistance and zero diameter, a very neat trick indeed. To simplify the problem, the current in the dipole element is assumed to be sinusoidal at the frequency of interest.

---

*When sinusoidal currents flow in a capacitor or inductor, energy is stored in the component at the peaks of voltage or current, which occur twice per cycle.

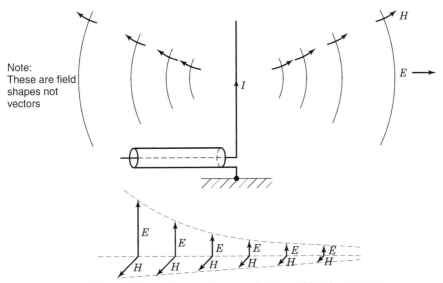

Relative field intensities as a function of distance from the antenna

**FIGURE 4.1**   Half-dipole antenna.

The half-dipole antenna as shown in Figure 4.1 is driven from a transmission line connected to the root or grounded end of the antenna. Each element of antenna current contributes to the radiation of the total antenna. The field around the antenna is current flowing in the capacitances of space. This goes back to the statement that a changing $E$ field is a current in space. This changing current further implies an $H$ field. Note that the field energy flowing out of the transmission line flows onto the antenna. In the space around the antenna, both the $E$ and $H$ fields intensity are changing as a function of height and radial distance. At the tip of the antenna, the current flow is zero. In the analysis the current pattern follows a half sine wave from the root of the antenna to the tip.

The energy in this antenna field is leaving and returning to the antenna at the same time. There is a net transfer of energy into space that does not return. Theoretically, Poynting's vector can be used to determine the net energy that flows past any surface area that surrounds the antenna. If there is a good impedance match at the end of the transmission line, no energy will reflect back on the transmission line. Matching the transmission line to the antenna is a major consideration in antenna design.

The current flow in the antenna causes an $E$ and $H$ field that diminishes in intensity with distance from the antenna. The ratio of the $E$-field intensity to the $H$-field intensity far from the antenna is a constant. This constant is known as the *impedance of free space*. The ratio of $E$ to $H$ is 377 $\Omega$ when the waves have little curvature. Near the antenna, the ratio of $E$ to $H$ is dominated by the $E$ field. This type of field is said to be a high-impedance electric field.

A distance of $\lambda/2\pi$ from the antenna is called the *near-field/far-field inter-face*. At 1 MHz the wavelength $\lambda$ is 300 m. This is the distance a wave travels in free space at the speed of light for 1 $\mu$s (the speed of light is approximately 300,000,000 m/s). Beyond 50 m the ratio of $E$ to $H$ is 377 $\Omega$. Beyond this distance both fields fall off linearly with distance and the radiation takes on a character known as a *plane wave* even though there is always some curvature to the wave.

Assume that the interface distance is 100 m and that the $E$ field at this point is 10 V/m; then the field strength is 5 V/m at 200 m. At this distance the $H$ field is 5/377 A/m. Beyond the interface distance the wave is always considered a plane wave. Inside the interface distance the ratio of $E$ field to $H$ field increases linearly. Near the dipole the field is said to be a high-impedance $E$ field. Very close to the antenna the $E$ and $H$ field intensities are complex in nature and the ratio of $E$ to $H$ does not follow a simple set of rules.

The $E$ field inside the near-field/far-field interface can be calculated as fol-lows. Assume that the interface distance is 100 m and the $E$ field is 10 V/m at this distance. At 50 m the wave impedance is doubled. The power cross-ing the two spherical surfaces must be equal. This requires the $E$ field must increase by a factor of $\sqrt{2}$ and the $H$ field must decrease by a factor $\sqrt{2}$. At this distance the ratio of $E$ to $H$ is $377 \times 2$ $\Omega$.

## 4.8   CURRENT LOOP RADIATORS

In most electronic applications, currents flow in small loops or small rectan-gles. In digital circuits there is a loop formed by the connections from a logic switch to the next gate. There is a loop from the power supply capacitor to a logic switch. Every loop carrying a changing current is a form of radiating antenna. The difference here is that the small current elements that form the antenna are arranged in closed paths rather than extending out onto a straight conductor or half-dipole. A loop radiator is shown in Figure 4.2.

There is a changing $E$ and $H$ field near any loop of wire carrying a si-nusoidal current. Ampère's law describes the $H$ field in the vicinity of the coil. Near the loop this $H$ field dominates. This radiating geometry also has a near-field/far-field interface distance which is again $\lambda/2\pi$, where $\lambda$ is the wavelength. ($\lambda = 300$ m at 1 MHz). For distances greater than this interface distance the waves are again called *plane waves*. Beyond this distance the ratio of the $E$-field intensity to the $H$-field intensity is constant and equal to 377 $\Omega$. Beyond this distance both the $E$ and $H$ fields fall off linearly with distance. Inside the interface distance the ratio of $E$ to $H$ decreases linearly. Near the current loop the wave impedance is low and the field is called a *near induction field*.

The $E$-field radiation from a loop of wire carrying a sinuoidal current is proportional to the loop area, the square of frequency, and inverse to distance.

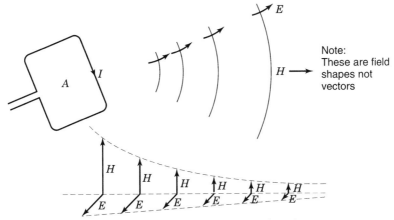

Relative field intensities as a function of distance from the antenna

**FIGURE 4.2** Loop radiator.

If the radiation level is known for one geometry, it is simple to estimate the radiation from another geometry. If many loops are involved, such as on a logic card, a worst-case analysis assumes that there is an equal contribution from each loop. A good starting point is the field strength at a distance of 1 m from a 1-cm$^2$ loop at a frequency of 100 MHz. The impedance of the loop is 30 $\Omega$ and the voltage exciting the loop is 1 V. The $E$-field level for this configuration is 316 $\mu$V/m.

To calculate the $E$-field strength for other geometries, use the following rules:

1. For greater distances, divide by the ratio of distances. Field reduces.
2. For greater loop areas, multiply by the ratio of areas. Field increases.
3. For greater voltages, multiply by the ratio of voltages. Field increases.
4. For higher frequencies, multiply by the ratio of frequencies squared. Field increases.
5. For more loops, multiply by the number of loops. Field increases.
6. For a higher-impedance load, divide by the ratio of impedances. Field decreases.

This is obviously a worst-case calculation.

Fields related to utility power are often *near induction fields*. The magnetic field near a power transformer is such an induction field. The near-field/far-field interface is more than 200 miles away. The electric shield that was discussed in Chapter 1 is very ineffective against a near induction field. More will be said later about shielding this field.

## 4.9   FIELD ENERGY IN SPACE

There are no field energy losses in space. The total energy crossing any spherical surface that surrounds a radiator is constant. Consider a radiator at the center of a sphere. For any small area on the surface of this sphere, the energy crossing that area per second is the product of the $E$ and $H$ fields times the area. If the power is radiated equally in all directions, the total power equals $E$ times $H$ times the area of the sphere or $P = \text{area} \times E \times H$. Since $E$ and $H$ are related by 377 $\Omega$, a little bit of algebra shows that $E = \sqrt{30P}/r$, where $P$ is in watts and $r$ is in meters. This illustrates the point that the $E$ field falls off linearly with distance in the far field.

Most antennas are designed to be directional. Television transmitting antennas are designed to direct energy to the city and not toward uninhabited areas. It is also wasteful to send energy toward the sky. A radar antenna is designed to send a ray of energy toward a target. Radar signals must be very intense, as a very small percentage of the radiated energy is reflected and returned to the radar set.

When a radiated signal is detected, an observer cannot easily determine the antenna radiation pattern. If an observer assumes that the antenna radiated the same energy in all directions, it is simple to calculate the assumed total radiated power from the measured field strength if the distance to the radiator is known. The ratio of apparent power to the actual transmitted power is known as *antenna gain*. In a radar transmission, the field strength at the target might be 10 V/m. At a distance of 10 km, this represents a radiation power level of 333 MW. If the antenna gain is 1000 and the radar bursts occur 1% of the time, the average power is only 3.33 kW, a practical power level.

## 4.10   PROBLEMS

1. A plane wave has an $E$ field of 20 V/m. What is the $H$ field?

2. The interface distance for a transmitter is 100 m. What is the frequency of the transmitter?

3. In the far field, the $H$ field 20 m from the radiator is 2 A/m. What is the $H$ field strength 30 m from the radiator?

4. What is the interface distance for a 2-MHz radiator?

5. A current loop radiates at 10 MHz. What is the interface distance? What is the wave impedance at half this distance?

6. A half-wave dipole radiates at 100 MHz. The $E$ field at the interface distance is 10 V/m. What is the $E$ field at half this distance? What is the wave impedance at this distance?

**7.** In problem 6, what is the $H$-field strength at this distance?

**8.** A radiating antenna has a gain of 10. The real power delivered to the radiator is 5000 W. What is the $E$-field strength at 20 km from the antenna? Assume that the measurement is made in the direction of best transmission. What is the field strength at 40 km?

## 4.11   REFLECTION

Electromagnetic waves are reflected at a conductive surface. If the surface is a perfect conductor, the reflection is total. In practice a small percentage of the wave enters the conductor and is lost in heat. If the conducting surface is very thin, some wave energy can actually pass through the conductor and continue its path on the other side. An electromagnetic wave that reflects from a conductive surface behaves much like a wave traveling along a transmission line. In a transmission line, the reflection from a short circuit is complete. If a transmission line is terminated in a low-value resistor, most of the energy is reflected and some of the energy is dissipated in the termination. This is the situation when a plane wave reflects from a practical surface. Most of the energy is reflected and a fraction is left to enter into the conductor. At the surface of the perfect conductor the $E$ field must be zero, which means that the reflected wave has a reversed $E$-field sign. For a practical conductor there is a net $E$ field at the surface. This represents a fraction of the incident wave that enters the conductor. There is an $H$ field for the arriving, reflecting, and penetrating wave at the conducting surface. This means that there is current flow on the surface.

The ideal reflection process assumes a very large conductive plane at right angles to the direction of the arriving wave. In practice, there are few large plane conductive surfaces and the wave energy does not arrive at right angles to the surface. The earth or the ocean are large conductive surfaces that are the exception. The reflections found in practical hardware situations are very complex and in most cases defy analysis. The only approach that makes sense is to consider a simple geometry and a worst-case scenario. Field measurements in or near hardware rarely offer a solution, as the very presence of a measuring device modifies the field strength. The circuit being affected by an interfering field is sometimes the only tool that can be applied to measure field strength.

The ratio of the reflected wave to the penetrating wave can be approximated by the ratio of four times the barrier impedance to the wave impedance. For example, the wave impedance of free space is 377 $\Omega$ [per square (□)] and the barrier impedance of a copper sheet can be 500 $\mu\Omega/\square$. The ratio of impedances $\times 4$ is 6127. This means that 1 part in 6000 (approximately) of the initial wave is not reflected.

## 4.12   SKIN EFFECT

The penetration of field energy into a conductor depends on frequency, dielectric constant, conductivity, and permeability. It is convenient to describe this penetration in terms of a skin depth. A skin depth is where the $E$ field is reduced to 36.8% of its surface value. In two skin depths the $E$ field is reduced by anther 36.8% or down to 13.6%. One skin depth for copper at 1 MHz is 0.0062 cm. The skin depth is proportional to the inverse square root of three factors: frequency, conductivity, and permeability. For example, at 10 MHz the skin depth of copper is 0.0020 cm.

The skin depth of iron can be calculated by considering the ratios of both conductivity and permeability to copper. Unfortunately, permeability falls off with frequency, and this complicates the situation. Note that the relative permeability of copper is unity.

When a current flows in a wire, there is field in and around the wire. In this case there is no plane wave reflecting off the surface. The field that penetrates the wire is the field that causes current flow. Any changing current results in a changing magnetic field, which by Lenz's law generates a counter voltage inside the conductor. This counter voltage limits the current flow. The result is an attenuation of current with depth. This reduction of current flow is also called *skin effect*. In most cases the skin depth can be approximated by the same equations that are used to describe the penetration of plane waves into large conducting planes. Skin depth is again a function of conductivity, permeability, and dielectric constant. A copper wire has a skin depth of 0.0062 cm at 1 MHz. It is thus very obvious that most of the current stays on the outside surface and the center of the conductor is not used. This has the effect of raising the resistance of the wire. The magnetic flux per unit current is essentially the same, so that the inductance per unit length is not affected significantly.

At 60 Hz, copper has a skin depth of 0.80 cm. This means that the resistance of large diameter power conductors is affected by skin effect. One device that is used in long power transmission lines is to make the core of the conductors out of steel. This increases the strength of the wire allowing support towers to be spaced farther apart. This is cost effective since skin effect limits the current to the outer portion of the conductor.

## 4.13   PROBLEMS

1. At 4 MHz, 99% of a plane wave is reflected at a copper surface. The initial wave is 10 V/m. What is one skin depth? What is the $E$ field at this depth?

2. What is the $E$ field at two skin depths for problem 1?

3. For 400-Hz power, what is one skin depth in copper?

**4.** The resistivity of aluminum is 2.83 $\mu\Omega$-cm. What is one skin depth at 60 Hz? At 400 Hz? At 1 MHz?

**5.** What is the skin depth for an iron wire at 1 MHz? Assume a permeability of 100 and a resistivity of 10 $\mu\Omega$-cm.

## 4.14  SURFACE CURRENTS

Fixed voltages between conductors result in surface charges on these conductors. If there is a steady (dc) current flow in a wire, some of the charges flow uniformly throughout the cross section of the wire. For a transmission line over a ground plane the current in the ground plane concentrates under the transmission line. The current flow patterns are controlled by field geometries. The depth of penetration is controlled by frequency and resistivity.

When a switch connects a fixed voltage to a transmission line, charges begin to flow out on the surface of the conductors. When the electromagnetic wave first reaches a fixed point along the transmission line, the charges start moving on the surface. As time progresses the current begins to penetrate the wire. After a suitable length of time the current flows in the entire cross section. This slow penetration of current into the conductor is the result of a counter voltage generated by magnetic flux entering the conductor. This is, broadly speaking, skin effect phenomena where the penetration changes with time. For most circuit geometries the transition from surface flow to midconductor flow occurs within the first few milliseconds.

The resistance of a square of conductive material is not dependent on the size of the square as long as the current flows uniformly across the square (see Section 4.4). The magnetic field associated with this current flow is greatest just below and above the surface. The resistance of a conducting square is very low, but connections to the square and any associated transmission line impedances must be considered. A point contact with the square concentrates the current, thus intensifying the magnetic field. This adds inductance and resistance to the current path. To avoid this concentration of current, many parallel connections should be made.

The resistance per square of a copper sheet 0.01 cm thick is 172 $\mu\Omega$. The skin depth at 1 MHz is 0.0062 cm. This means that a thin sheet of copper such as a plated ground on a printed circuit board has about the same low resistance per square from dc to about 1 MHz. (Up to this frequency the current flows in the entire cross section of the sheet. Note that the sheet is thinner than one skin depth.) Above 1 MHz the ohms per square for this copper sheet rises as the square root of frequency. For this example the value rises to 543 $\mu\Omega/\square$ at 10 MHz. In many applications a thick conductive surface is used just to make the surface robust and manufacturable. Used correctly, thin sheets of conductive material can provide a very low impedance over a wide frequency range. This is not true for simple round conductors where the inductance is proportional to length. Again—it's not the amount of copper that is used, but the geometry of that copper.

## 4.15   GROUND PLANES AND FIELDS

The term *ground plane* can mean anything from the surface of the earth to a layer of copper on a printed circuit board. It is accepted practice to provide one or more ground planes on circuit boards used to mount high-speed digital components. These conductive surfaces together with interconnecting traces form short transmission lines. These traces are used to transport digital signals between logic components. The traces are a few thousandths of an inch from the ground plane, which concentrates signal fields in a small volume of space. The current that flows for each signal is returned in the ground plane area immediately below the trace. If there are conductive planes above and below the traces, current flows on both planes. This geometry confines the fields to the volume between conductive planes, and this significantly limits radiation from the circuit.

Ground planes are often used in facilities housing many racks of electronic hardware. These planes may be sheets of metal, grids of steel mesh, or metal stringers resting on steel stanchions. Cabling and power distribution are usually placed on the floor under these structures. Used correctly, these surfaces can provide lightning protection and electrostatic discharge (ESD) control. These surfaces can reflect electromagntic field energy but they cannot absorb this energy. Conductive grids do allow some fields penetration at high frequencies. More will be said about this penetration in the next section. To approximate the quality of a solid conductive surface, the bonding at the grid intersections must be of high quality. The ohms-per-square value of a solid ground plane is in microhms. This means that the bond between stringers should also be in the same microhm range if the plane is to be effective. This requires plated surfaces that are held together under pressure.

## 4.16   APERTURES

An *aperture* is an opening in a conductive surface. It can be a ventilation hole, a seam where two pieces of metal join together, or the mounting hole for an electronic control or one of the holes in a wire mesh. When an electromagnetic wave reflects off a surface containing an aperture, some of the field can penetrate the aperture. An example might be a window in a building wall, which allows sunlight to enter. Even if the hole is small, the light intensity in the resulting beam is still maximum.

The reflected wave causes surface currents to flow on the conductive surface. This pattern of current flow is modified by the presence of an aperture. The $E$ field is zero on the conducting surface except in the area of the opening. The question is: What is the $E$ field on the other side of the aperture?

The problem of calculating the field that penetrates through an aperture is very complex. Consider a very idealized problem. Let the conducting surface have zero resistivity and cover an infinite area. Let the plane wave that reflects from the surface arrive at right angles to the surface. Further let the conducting material be very thin and have only one aperture, a round hole. The wave that emerges from the other side is no longer a plane wave, and there are surface currents on this far side. Even if this problem had a simple solution, it would be difficult to use the result. Many apertures are rectangles and these field solutions are almost impossible to handle. For example, the field intensity near the corners has a great deal of fine structure, which is of little practical importance.

In most applications the energy that enters through an aperture is inside a conductive enclosure with associated electronics. If the aperture problem is difficult, the problem of determining field strength inside a real enclosure is not solvable. The only practical way to determine field strength might be to note the impact on a specific circuit. One approach that is used is that of *worst-case calculation* (WCC). The assumptions are as follows:

1. Assume that the external wave arrives perpendicular to every surface.
2. Assume that the $E$ field is aligned with the maximum dimension of the aperture.
3. Assume a thin perfect conducting surface.
4. Assume that the wave on the other side of the aperture is a plane wave.
5. Assume that no internal reflections or absorptions occur.
6. Assume that the wave is not attenuated when the aperture has a maximum dimension greater than one-half wavelength.
7. Assume that the wave is attenuated by the ratio of maximum aperture dimension to the half-wavelength of the entering field.

The idea here is that the field energy that enters through the aperture is guaranteed to be less than this WCC. If this level of field strength poses no difficulty, there is no problem to solve.

## 4.17 MULTIPLE APERTURES

When several apertures exist on a product, the WCC assumes that each aperture contributes to field penetration. Even if the apertures are on opposite sides of the box, each aperture adds to the field penetration. If the external field strength is 10 V/m and the first aperture attenuates the field by a factor of 100 and the second by a factor of 50, the WCC assumes that the resulting field is 0.1 V/m + 0.2 V/m, or 0.3 V/m. If there are numerous apertures, the field intensity that penetrates can never be larger than the initial field intensity.

When apertures are close together, the story changes. Examples might be ventilation holes, wire meshes, or even seams that are riveted or screwed together. The penetration of field requires that there be free current circulation around each aperture. In the case of vent holes, wire meshes, or seams, the current cannot flow freely around each aperture. In these cases the WCC allows the group of apertures to be reduced to that of one aperture.

The issue often arises of how to treat a seam that is seemingly lighttight. If there is no continuous metallic bond, a WCC says that an aperture still exists. The maximum aperture dimension is the length of the seam. Even machined metal surfaces that appear to touch are suspect. To close this type of seam, a metallic gasket must be used to bond the surfaces together. The reason relates to the surface resistance, which must be microhms per square across the seam. Machined surfaces provide many point contacts, which are not sufficient to provide for a smooth current path on the surface. Surface currents must travel into the interior of the conductors to effect the connection. As a result, field energy can penetrate such a seam. This penetration can occur into or out of an enclosure.

## 4.18  WAVEGUIDES

A *waveguide* is a conductive cylinder (often, rectangular) without a center conductor. A waveguide will allow an electromagnetic wave to propagate provided that the wavelength is short enough. At 4 GHz a half-wavelength is 3.75 cm. This is approximately the lowest-frequency wave that can propagate down a cylinder of this size. There are a variety of patterns associated with microwave propagation, which means that there are quite a few frequencies that will propagate down the waveguide just above 4 GHz. At higher frequencies the specific frequencies that can propagate are fairly close together. The WCC assumes that all frequencies above the critical frequency will propagate freely in the conduit. Below the critical frequency all propagation is attenuated very rapidly. The attenuation depends only on the ratio of waveguide depth to the width of the guide. If the ratio is $2:1$, the attenuation factor is 1000. If the ratio is $4:1$, the attenuation factor is 1,000,000. Obviously, an aperture built as a waveguide is a very powerful tool.

A waveguide with a center conductor is a transmission line valid at all frequencies. A small tube of metal can be a very effective way to bring a fiber optic cable into a piece of hardware. Any steel support wire must be removed if the waveguide attenuation is to be realized.

## 4.19  ATTENUATION OF FIELDS BY A CONDUCTIVE ENCLOSURE

The WCC of the attenuation of an arriving field for a conductive enclosure can now be considered.

1. Field energy is reflected at the surface. The percentage that is not reflected enters the surface and is attenuated by skin effect. If the skin is

thin enough, some of the energy enters the enclosure. In a WCC this field strength is independent of the areas or shapes of the walls.

2. Each isolated (independent) aperture allows the field to penetrate the enclosure. The field attenuation is the ratio of half-wavelength to the maximum aperture dimension. Each aperture contributes to field penetration. Apertures that are close together (dependent) act as one aperture, which allows field energy to penetrate the enclosure.

3. Waveguides function as follows: The opening attenuates the field just like any other aperture. The depth of the aperture further attenuates the field strength based on the length-to-width ratio of the waveguide. This assumes that the frequency of interest is below the lowest waveguide frequency. This is called the *wavelength beyond cutoff*.

4. The total field in the enclosure is the sum of the fields from skin penetration, aperture penetration, and waveguide penetration.

## 4.20 GASKETS

Gaskets are often used to close apertures. Typically, gaskets are made of conductive material that makes contact with the mating surfaces. The surfaces in question must be free of paint or anodization. To guard against rust, the surfaces are often plated. Gaskets often contain stainless steel filaments that dig into the contact surfaces. A wide range of products are commercially available that fit the needs of many applications. Gaskets that conform to irregular surfaces must be replaced if they are ever removed.

It is interesting to consider the use of screws to reduce the dimensions of an open seam. To reduce the field penetration by a factor of 10, the number of screws need to be increased by a factor of 10. This tight spacing turns out to be very impractical. It makes more sense to design the seam with an overlap so that waveguide attenuation is available. If the ratio of aperture width to depth is a factor of 2 or 3, the field attenuation will be significant.

## 4.21 HONEYCOMBS

The need for ventilation and high-field attenuation can be met using a honeycomb construction. A honeycomb consists of many small conductive hexagonal tubes bonded together on all exterior surfaces. A group of 100 or so tubes can provide for air passage. Note that current can flow freely around the inside surface of each tube. This means that each tube is an independent aperture. The high-field attenuation is the result of having depth to each aperture.

Consider a 100-tube honeycomb that has a tube dimension of 1 cm. If the tube length is 4 cm, the wave guide attenuation is 1 million to 1. Consider an *E* field of 10 V/m at 100 MHz. The half-wavelength is 75 cm. The *E* field that enters each aperture is 10 V/m × the ratio 1/75, or 0.133 V/m. The field strength at the end of each aperture is 0.133 $\mu$V/m. Since there are 100 independent apertures, the field that enters the enclosure is 100 times greater, or 13.3 $\mu$V/m.

The honeycomb must be bonded to the conductive surface around its perimeter. This bonding is usually provided by a gasket. If the bond has a 1-cm gap, the field energy that enters this gap is 1/75 times 10 V/m, or 0.133 V/m. This leakage completely negates the effectivity of the honeycomb. For this reason, honeycombs that are commercially available are supplied with a gasket arrangement. This gasket must be applied over a prepared surface to guarantee performance.

## 4.22  WAVE COUPLING INTO CIRCUITS

Plane waves that are moving in space are modified by any set of conductors. For plane conductive surfaces there are reflections and resulting surface currents. For cables and circuit conductors the loop areas involved play a key role. Remember that the energy flowing in space always tries to get to a lower-energy state. The impedance of free space is 377 $\Omega$, which makes it easy for the field to follow a new path between pairs of conductors. These new paths often lead directly into equipment. The field patterns are never simple and a WCC is needed to estimate the nature of the coupling.

For near induction fields it is best to use the $B$ field to calculate the voltage induced in any circuit loop. The $H$ field is first determined by Ampère's law. The $B$ field is then equal to $H$ times the permeability of free space. The flux that is captured by a loop is the $B$ field times the coupling area. The voltage induced in the loop is the rate of change of the flux.

For plane waves it is simple to use the $E$ field to approximate the voltage induced in a conductive open loop. Figure 4.3 shows an input cable over a ground plane. The $E$ field is vertical and the wave travels parallel to the cable. The loop in question is formed by the cable and the ground plane. The voltage in question is at the ungrounded end of the cable. At a moment in time the $E$ field intensity varies along the loop. The voltage across the loop is simply the $E$ field times the spacing. Let the cable length be one half-wavelength. When the $E$ field is maximum positive at one end of the cable, it is maximum negative at the other end. The maximum voltage induced in the loop is simply twice the $E$-field intensity times the spacing. Assume that a 10-MHz $E$ field has a level of 20 V/m. The cable length is 15 m. The cable is spaced 10 cm over the ground plane. The induced voltage is twice the 20 V/m × 0.1 m, or 4 V rms.

For cable lengths less than 15 m, the induced voltage is approximately the maximum voltage times the ratio of length to half-wavelength. For example, if the cable length is 7.5 m, the induced voltage would be 2 V. In theory, when the dimensions exceed one half-wavelength, the induced voltage is less. In the

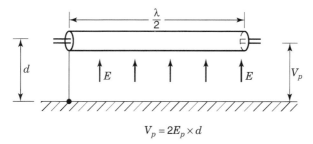

$$V_p = 2E_p \times d$$

**FIGURE 4.3**  Cable over a ground plane.

spirit of WCC, when one of the dimensions exceeds one half-wavelength, the half-wavelength number is used.

The field coupling is thus proportional to both cable length and height above the ground plane. This means that the coupling is proportional to loop area. This is a very important point to remember. To reduce coupling, reduce the coupling loop area. Field coupling occurs in all conductive loops not just those formed over a ground planes.

## 4.23  PROBLEMS

1. A metal box has a seam closed by screws that are spaced 10 cm apart. There are 20 screws around the perimeter. There are a group of ventilation holes in a close pattern. There are 15 holes each $1 \times 3$ cm. There is a meter opening that is $6 \times 8$ cm. Determine the WCC for the $E$ field inside the box if the external field strength is 10 V/m at 20 MHz.

2. A honeycomb has 50 openings that are 1.5 cm across and 6 cm long. What is WCC for the attenuation of an external field at 100 MHz?

3. A box has three isolated openings. Their dimensions are 2, 3, and 4 cm in diameter. A wire screen covers another opening that is $10 \times 20$ cm. The screen openings are 5 mm. What is the WCC for field penetration for a $-2$ V/m plane wave at 10 MHz?

4. A cable runs over a ground plane at an average height of 20 cm. What is the voltage coupled to the loop if the cable run is 10 m? Calculate the WCC at 1, 10, and 100 MHz if the field strength is 20 V/m.

## 4.24  SQUARE WAVES

Data or information in electrical form surrounds us. The signals driving a loudspeaker or a video pattern are examples of such electrical activity. In our computers, digital signals that represent words, numbers, or instructions are sequences of two voltage or current levels (ones and zeros). All of these signals vary in time. These variations represent information flow in one way or another. The ability to move information is related to how rapidly a circuit can respond to time-varying signals.

Circuits are often analyzed using sine waves. A sine wave represents three pieces of information. The first is amplitude, the second is frequency, and the third might be phase. When testing the response of a circuit with a single sine wave, very limited information results. To get more information, many different sine waves need to be used. The sine waves might vary in frequency and amplitude. Testing using sine waves at many different frequencies is time consuming.

A square wave is made up of many sine waves. A *square wave* is a voltage or a current that has two signal levels and repeats over and over again. When a square wave is used to test a circuit, the information content is much greater. In fact, the response to a square wave can imply what the response will be to a wide band of sine waves. A square wave has a set of characteristics defined in time. These characteristics are the rise time, voltage level, hold time, and fall time. A linear circuit responds separately to each of the sine waves that make up the square wave. The result is the sum of these individual responses. The question is: How do we relate the square-wave parameters in time to frequency parameters?

When signals are repetitive, such as square waves, sawtooth waves, or triangle waves, a group of sine waves can be summed to duplicate these signals. The amplitudes of these sine waves can be determined by a Fourier analysis, a discussion of which topic is beyond the scope of this book. The sine waves that make up a perfect square wave are discussed below.

Suppose that a repetitive square wave is symmetric about the voltage axis and starts at time $t = 0$ and the voltage level changes every $\frac{1}{2}$ ms. The voltage is 1 V for $\frac{1}{2}$ ms, then $-1$ V for the next $\frac{1}{2}$ ms. In the second millisecond the voltage again changes from plus to minus 1 V. The period for this pattern is 1 ms. This ongoing waveform is known as a 1-kHz square wave. The first sine wave that makes up this square wave (also called the *fundamental*) has a period of 1 ms or a frequency $f$ of 1 kHz. The frequency $f$ of the second sine wave is 3 kHz and the amplitude is one-third of the fundamental. The third sine wave is 5 kHz at an amplitude of one-fifth of the fundamental. The sine waves at $3f$, $5f$, $7f$, and so on, when summed, make up a square wave. The sine waves at $3f$, $5f$, $7f$, and so on, are all called *harmonics*. The sum of the first few harmonics are shown in Figure 4.4. This symmetrical square wave is made up of odd harmonics only. The peak amplitude of the fundamental sine wave must be set to 0.785 V, so that the sum of all the harmonics yields a 2-V peak-to-peak square wave.

At the moment when a square wave is first connected to a passive (*RLC*) circuit, none of the capacitors have charges and the current in all inductors is zero. After a transitional or transient time period the current and voltage

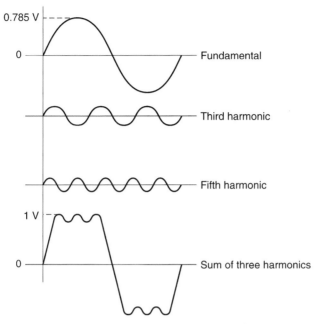

**FIGURE 4.4**   Sum of the first few harmonics that make up a square wave.

waveforms for every component repeat each cycle. This is called the *steady-state condition* of the circuit. After the transient period is over, the current flow or a voltage drop associated with every component will not be zero when the square wave makes a transition.

The rise time of any practical square wave is finite. This is the time it takes the voltage or current to change between levels. Square waves can be produced that change levels in nanoseconds. All that is necessary in testing is that the square wave change levels in a time shorter than the transition time of the circuit being tested.

When square waves are not symmetrical about the zero time axis, the harmonic content of the square wave is shifted in phase. If the average value of the square wave is not zero, it is said to have a *dc component*. If the square wave is not symmetric about the amplitude axis, even harmonics must be a part of the waveform makeup. For a finite rise time the higher harmonics are reduced in amplitude. The exact nature of the harmonic content for standard waveforms is well covered in the literature.

When a square wave is used to test a circuit, the response represents the effect of many simultaneous sine waves. If the circuit responds with ringing (high-frequency content near the leading edge) or a large overshoot, this is often an indication of instability. If the resulting signal sags in time, this may be an indication of poor low-frequency response. All of this information comes from a single observation. This is the power of a square wave.

## 4.25  HARMONIC CONTENT IN UTILITY POWER

Electronic loads are often nonlinear. For example, capacitors used in rectifier systems demand current at the peaks of voltage. Line current may flow for less than 25% of the power cycle. Transformers that draw excess magnetizing current demand line current at the zero crossings of voltage. (This is when the flux level is at its maximum.) These peaks of current flow in the resistance of the utility line and in the resistance of the transformer coils. Any resulting voltage drop distorts the voltage waveform at the transformer. Light dimmers and motor speed controllers work by connecting to the power line over limited portions of the power cycle. The resulting current waveforms can be analyzed in terms of harmonics, the same as any voltage waveform. Variations in power line voltage waveform are apt to be most severe on the power line near the equipment in question. Generally, the voltage waveform improves at the service entrance and is improved further at the distribution transformer.

Motors can take current from the power line in a nonlinear manner. Induction motors with a slip frequency can add line voltage content at a frequency lower than the power frequency. Switching regulators take power in short pulses. In some designs the pulse duration varies as the amplitude of the power line voltage changes. All of these loads are nonlinear and can modify the line voltage waveform.

There are many sources of power line waveform distortion. If these distortions repeat each cycle, they produce harmonic content at frequencies that are multiples of the power fundamental. The higher the harmonic frequency, the lower the capacitive reactance through parasitic or mutual capacitances in the power transformer. In general the higher the harmonic frequency, the lower the amplitude of the harmonic. This means that the resulting parasitic current flow pattern roughly reconstructs the waveform of the voltage distortion. When this current flows in a secondary circuit conductor, the interference that is introduced has the same waveform as the power line voltage distortion. Shields in transformers that supply electronic loads can reduce this parasitic current flow. These shields are discussed in Section 3.18.

Nonsinusoidal line currents imply fields around the conductors that are rich in harmonics. These higher-frequency components are often seen in interference patterns, as they couple more easily to nearby circuits. Switching devices on the line are perhaps the worst offenders, as the current waveforms can have short transition times. The energy that is demanded must come from the power line (transmission line) fields. These rapidly changing field patterns can be sensed by other devices connected to the same utility power. This interference can also couple to nearby signal lines, conduit, and building steel. Once the genie is out of the bottle, it is difficult to put him back. If energy is needed in short bursts, it is preferable to supply this energy locally from a line filter. This is discussed in Section 5.22.

## 4.26   SPIKES AND PULSES

Spikes or short pulses of voltage appear frequently on power lines. These pulses vary in duration and amplitude and can occur at any point on the power cycle. These pulses can arise from power and load switching. Pulses can do damage to hardware and cause errors in digital processes. Pulses can also result from an electrostatic discharge or from a lightning stroke. Obviously, pulses are not limited to power lines. A voltage pulse has an associated changing $E$ and $H$ field. This field propagates between conductors and can easily cross-couple to other circuits.

A pulse is a singular electrical event. It has rise time, fall time, amplitude, and duration. Repetitive waveforms allow for a harmonic analysis. Single events do not lend themselves to this approach. It can be shown that a single voltage pulse can be generated mathematically by adding together sine waves at all frequencies where there is an amplitude and phase relationship. These sine waves start at a very early point in time and at the expected moment their sum generates a pulse. After that time the signals continue to run with their sum (average value) equal to zero. The amplitude of any one sine wave at any one frequency is zero. There is a finite amplitude for any selected band of these sine waves. The mathematics behind this approach is obviously very sophisticated.

The actual response of a circuit to a single pulse is beyond the scope of this book. Usually, the information that is needed is the peak amplitude of the response. The details of the response waveform are not important. Amplitude information is needed to decide if there is a chance of overload, damage, or error for the circuit in question.

A simple way to analyze the peak response of a circuit is to substitute one continuous sine-wave for the pulse. The resulting sine-wave amplitude is an approximation to the amplitude response for the actual pulse. The frequency that is selected is given by $1/\pi\tau_r$, where $\tau_r$ is the time it takes the pulse to transition from 10% to 90% of final value. The peak amplitude of the test sine wave is the voltage difference between the 10% and 90% points. This technique may seem crude, but the method provides insight into a very difficult problem. This analysis has a built-in safety factor. The exact answer does not matter, as the issues of overload, damage, or interference are rather subjective anyway. If circuit damage occurs at 6 V, it is prudent to design for a much lower peak response. In critical designs the accepted limit might be 0.6 V. This is a safe margin for this WCC analysis.

This rise-time approach to analysis is not limited to pulses. The maximum response to any repetitive waveform can be approximated by using this method. The shortest rise or fall time determines the frequency, which is $1/\tau_r$. The peak amplitude of the sine wave is the difference between the 10% and 90% points. A single sine wave is used to drive the circuit or system. The amplitude response will be an indicator of the peak response to the initial waveform.

## 4.27   TRANSFORMERS

The transformer action discussed in Section 2.18 assumed very ideal core material. The permeability of the core material was held constant and there was no discussion of core losses. In practice, the relationship between the $B$ and $H$ fields in magnetic materials is rather complex. The magnetic field that crosses the conductive core causes current flow that heats the core. Further, there is some magnetic flux that does not couple the primary and secondary coils. This flux is said to store energy in the leakage inductance of the transformer.

A plot of the $B$ and $H$ flux in transformer iron forms a hysteresis curve. A typical plot is shown in Figure 4.5. Ideally, the $B$ flux level follows a sine wave in time. This means that the vertical axis is traversed sinusoidally. The $B–H$ curve is traversed counterclockwise. On the upper curve as the $B$ field increases toward maximum, the $H$ field begins to increase more rapidly. The value of $B$ is maximum when the sinusoidal voltage applied is making a zero crossing. After the applied voltage reverses in direction, the $B$-field intensity falls and the $H$ field (magnetizing current) reverses in polarity and also falls. The hystereses curve is traversed once per power cycle. The ratio of peak $B$ to peak $H$ is defined as the *effective permeability*.

If the applied sine-wave voltage is reduced in amplitude, the nature of the hysteresis curve changes. If the core was already near saturation, the permeability at reduced voltage levels might increase. If the core was not near saturation, a reduction in applied voltage might reduce the effective permeability. This is the case shown in Figure 4.5. Core manufacturers usually provide a curve of effective permeability plotted against peak values of $B$ for sine waves. Because of symmetry, manufacturers may only show the top half of the $B–H$ curve.

In an ideal transformer, if the impressed voltage is sinusoidal, the $B$ field is sinusoidal. The resulting $H$ field is represented by the magnetizing current (unloaded primary current), which peaks twice per cycle. For some core materials the magnetizing current can be a quasi square wave. In audio transformers the $B$ and $H$ flux traverses small minor loops as well as large loops. Core materials should be selected that optimize the effective permeability for these small flux traversals. A core material that loses permeability for small signals could distort the signal.

In small transformers the magnetizing current flowing in the primary coil resistance modifies the impressed voltage. If this current is nonlinear, the voltage across each turn of the primary is no longer sinusoidal. This means that now both the $B$ and $H$ fields are nonsinusoidal. The result is that the secondary coil voltages are nonsinusoidal. If the secondary supplies nonlinear load current (rectifiers), the resulting current flowing in the primary coil further modifies the $B$ field. In some applications, experimentation is the simplest way to arrive at a practical transformer design. Calculations that consider all the nonlinear effects are difficult to handle.

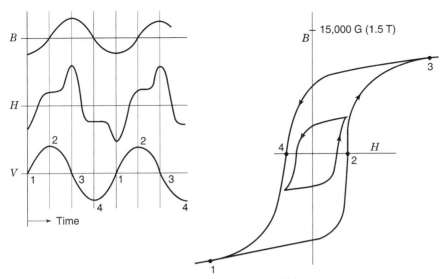

The $B$ field is sinusoidal when the applied voltage is sinusoidal.

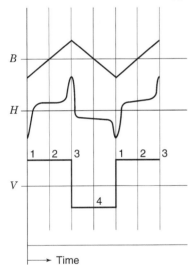

The $H$ field when the applied voltage is a square wave. The $B$ field is triangular.

**FIGURE 4.5**  Typical $B$–$H$ curve for transformer iron.

The majority of power transformers are operated at 60 Hz. When a transformer is designed to operate at a higher frequency, many parameters change. Assume that the $B$-field maximum at 60 Hz is 1.0 T. If all parameters remain constant, the $B$-field maximum at 400 Hz would be only 0.15 T. The $H$ field in the core would also be reduced by the same factor. It is obvious that this transformer could be designed with fewer turns on the same core. Fewer turns suggests a new design using a smaller transformer core. A smaller core implies a shorter magnetic path length, and this reduces the requirement for magnetizing current. Remember that the amount of magnetizing current depends on the magnetic path length and the number of turns.

Magnetizing current is needed to supply the $B$ field that supports the primary voltage. Without a magnetic core the transformer magnetizing current would be excessive. As the frequency rises the need to use magnetic materials to limit magnetizing current is reduced. At frequencies above about 300 kHz, it is practical to build transformers without a magnetic core. At these frequencies the leakage inductance and the parasitic coil capacitances severely limit the performance of the transformer.

## 4.28   EDDY CURRENTS

Magnetic core materials such as silicon steel that are used for power transformers are conductive. The magnetic field must cross this conductive material, and this results in current flow in the core. This current is called an *eddy current*. It is necessary to limit this current flow to avoid the obvious heating. To limit eddy current, transformer cores are usually made from stacked laminations. These laminations (thin layers of magnetic material) are stacked into the bobbin carrying the coils of wire. The laminations are interlaced so that a full stack does not fall apart. Common lamination shapes include E/I, E/E, D/I, or L/L. The E/I shape is very common. The lamination thickness for small 60-Hz power transformers is about 0.015 in. These laminations are coated with an insulating material, so the resulting core does not appear as a solid conductive block. The magnetic path for each layer is broken at the E–I interfaces. This technique adds some gap to the core but it keeps eddy currents from flowing between laminations. Interlacing the laminations allows the flux to cross over the gaps in adjacent laminations.

Power transformers at 400 Hz require laminations whose thickness might be 6 mils. These thin laminations are more difficult to handle. Very thin magnetic/conductive tapes can be used to laminate the core for applications above 400 Hz. Magnetic tape material is insulated on one side and usually wound in the form of a toroid.

Audio transformers must function over a wide frequency range. The core size is related to the voltages at the lowest frequencies. The flux swings at the higher frequencies are very small. This means that the lamination thickness does not need to accommodate high-frequency losses. In today's audio market, most audio transformer are replaced by solid-state circuitry. Audio transformers provide a tone quality that satisfies some aficionados, and they are still used in some more expensive equipment.

Power distribution transformers are often rated ($k$ factor) to accommodate high harmonic current flow. Typically, electronic loads are high in harmonic content, which increases transformer core and copper losses. The $k$ factor must be high enough so that these loads do not overheat the transformer.

## 4.29   FERRITE MATERIALS

Many power supply applications require transformers that operate above 50 kHz. At these frequencies the transformers are small and very effective. The cores that make up these transformers are made from powdered ferrites. The core is an alloy of magnetic materials in the form of a fine powder. This fine power is fired in a kiln with a ceramic filler. The resulting core material is an insulator. The core is made up of many small volumes of magnetic material. This limits core losses at these high frequencies. These cores can be used for inductors as well as transformers. In a typical application the coils are wound on a bobbin and the core in the shape of two cups fits through and around the coil bobbin. This surrounding core geometry limits external fields. The surfaces of the core that fit together are highly machined and polished to eliminate any gap. This is necessary if the core is to have an effective high permeability. In some core versions a precise air gap can be provided. These cores can be used to build inductors or transformers that require dc ampere turns. A gapped transformer can store energy that can be transferred to secondary circuits. This energy storage and transfer is at the heart of switching regulator design.

## 4.30   PROBLEMS

1. A ferrite core has a magnetic path length of 5 cm and a cross section of 1 cm$^2$. The effective permeability at 50 kHz is 5000. Assume a peak $B$ field of 0.1 T. How many turns are required to support a square wave at 20 V peak? What is the magnetizing current maximum? When in the cycle does this maximum occur?

2. A pulse rises to 100 V in 1 $\mu$s and returns to zero in 2 $\mu$s. Estimate the peak voltage coupled onto a 0.001-$\mu$F capacitor through a resistance of 300-$\Omega$ resistor. Use the approximation method involving sine waves.

3. What is the radiation at 30 MHz from a board with 100 loops with an area of 1 cm$^2$ and 100 loops of 2 cm$^2$? Assume a distance of 3 m and a logic level of 10 V. Assume that the impedance of the circuits is 500 $\Omega$.

4. A transformer supplies 100 W to a load, 10 W to core and wire losses, and it has a magnetizing inductance of 5 H. What is the primary current? Assume a sinusoidal magnetizing current and a primary voltage of 117 V at 60 Hz.

5. A ground return follows a loop around the perimeter of a circuit card. It carries a clock current that rises to 100 mA in 0.1 $\mu$s. The area of the ground return loop is 100 cm$^2$. What is the peak $H$ field 2 cm from the loop? What is the voltage induced in a 2-cm$^2$ loop that is 2 cm away from the loop? If the logic level is 5 V, estimate the radiation level at a distance of 3 m.

## 4.31   REVIEW

Conductors and their geometry control both the transmission and radiation of field energy. This energy can be reflected, absorbed, and coupled into circuits. Conductor geometries can be selected that limit radiation and/or coupling. Signals of interest and interference are both carried by fields. Conductor geometry determines where these fields can propagate. Keeping the paths separated is the key to good design.

Conductors carry currents on their surfaces at higher frequencies. Very thin conductive sheets can provide a low-impedance path provided that current flows uniformly in the sheet. Ground planes can help control fields, but they cannot attenuate them.

Transformers are an important part of the interference process. Energy can cross from the primary to the secondary coils and couple into signal circuits. A single shield provides limited protection. Magnetizing current and nonlinear loads can reflect into neutral conductor voltage drops, and this can be a source of interference.

A worst-case approach can be used to calculate the impact of radiation through apertures and waveguides. The effect of nonsinusoidal waveforms can be estimated by considering rise time and amplitude content.

# 5 Analog Design

## 5.1 INTRODUCTION

There is a definite feeling among designers that there is a field of analog design set apart from digital design. Digital design has as its goal a set of discrete logical steps that perform a set of functions. The design might consist of finding a suitable processor, memory, buffers, registers, logic gates, and perhaps some line drivers and using them as a circuit. A good part of the effort rests in firmware or software. Analog design has a different set of goals. Instead of just two logic states, all voltages or currents between two limits are important. Analog design involves topics such as filter theory, feedback theory, gain, phase character, input impedance, accuracy, linearity, and distortion. Analog and digital designs have many common problems. Both designs are concerned with rise times, delays, output currents, transmission processes, energy storage, and interference levels. In fact, at the very heart of digital design is analog design. Digital integrated circuits would not perform well if analog processes were not considered carefully. Digital design as pure logic is definitely separated from analog design.

When a transmission line processes a digital data stream, the process is analog. When a logic gate closes, the energy that must flow down the transmission line is taken from a decoupling capacitor, and this process is analog. When a clock signal couples improperly into an analog-to-digital converter, this is an analog problem. When logic signals are converted to light for optical transmission, this is an analog process. It is easy to see that analog processes abound in digital design.

Facilities that operate many pieces of electronic hardware are often exposed to interference problems. The difficulties are related to the large number of electronic devices, utility power distribution, and number of signal interconnections. Ground potential differences, radiation from hardware, crosscoupling from a common power source, long cable runs, exposure to lightning strikes, and electrostatic discharge can create problems for any type of facility. It is fair to say that these processes are analog in nature.

## 5.2 ANALOG SIGNALS

Before the digital revolution, most designs were analog. These designs revolved around audio equipment, radio receivers and transmitters, analog computers, and transducer instrumentation. The measurement of basic physical phenomena is still analog in nature. For example, analog measurements include temperature, pressure, displacement, strain, and velocity. Transducers are available that convert these parameters to voltages or currents. These signals are analog representations of actual physical conditions. It is often required to transport these analog signals to a central point for processing (conditioning), recording, or analysis. The selection of transducer type, cable type and conditioner are all part of analog design. A design objective is to handle these signals so that their integrity is maintained.

Many analog transducers produce small signals. Thermocouples can generate a few tens of millivolts. A pressure transducer might generate 10 or 20 mV full scale. If the intent is to maintain an error level of 0.1%, interference must be kept below a few microvolts. This is where there is a basic difference between analog and digital signals. Digital signals can tolerate an error of a few tenths of a volt and not change the logic level. A pulse on an analog line is an obvious error, but a pulse that changes a digital word might disrupt an entire computer.

The bandwidth requirements for analog signals can vary from dc to many megahertz. One reason to add filtering to the signal is to take out interference introduced by the environment. In most cases the filtering is used to remove higher-frequency content. Filters add delay to the signals, and this can be a problem if events are to be time related. It is preferable to avoid the use of filters unless the data are buried in extraneous noise. If the data are converted to a digital format, post digital filtering is a practical solution. Analog filtering along with bandwidth selection can be expensive. Generally, analog filtering is handled after the signal has been amplified. Care must always be taken that the interference does not exceed the limits of the electronics handling the signal. A filtered overloaded signal is totally invalid.

Analog or digital signals that are carried on conductive cables are associated with two ground reference potentials. The signal origin may be electronic hardware or a transducer. Even if there is no ohmic contact with a ground, there is a capacitance to this reference potential. The cable carrying the signal terminates on a second ground reference potential. At this second point there is usually hardware associated with conditioning the signal. Connecting the shield at both reference grounds does not eliminate the potential difference between grounds. The issue is cable length, shield current flow, and the nature of the fields in the environment. It is always prudent to accept the potential difference and design the electronics to reject the potential difference.

## 5.3   COMMON-MODE INTERFERENCE

In Sections 3.18 and 3.19 we considered the electrostatic shield around sensitive circuits. The rules for connecting the shield to the signal reference or common conductor were presented. A signal shield must be connected to the signal common at the point where the signal grounds (connects to an external reference conductor). No other shield connection is allowed. When a signal is sensed in one signal reference environment, conditioned (amplified), and connected to circuitry in a second reference environment (grounded), problems can result. Connecting a conductor such as a shield to both grounds is termed a *ground loop*. The connection in effect allows electromagnetic fields in the area to circulate current in this new path. This unwanted current is now very close to a sensitive signal path. This added current is apt to be power related. As a result of this current, the shield assumes a gradient of potential along its length. At the midpoint the shield will be at one-half the ground potential difference. This results because there is voltage drop in the inductance and resistance of the shield. For typical shields this gradient of potential couples to the shielded conductors. This coupling is a function of line balance, line impedance, and frequency as well as the cable characteristics. The resulting interference may not be acceptable in some applications. Grounding the shield at both ends can be unsatisfactory. The problems are most severe when the distances are large or the electromagnetic activity in the area is significant. Hardware might function well in the laboratory but fail in a practical installation. It takes only one significant pulse to damage some hardware.

Figure 5.1 shows a shielded pair of signal conductors where the signal common and the shield are connected to a reference ground at a remote point. This cable carries a signal of interest that is associated with the source ground. It is simple to say that there is a potential difference between the two grounds. A more accurate statement says that there is a voltage difference caused by fields that cross the loop formed by the cable and the ground connections (ground plane). This voltage is common to the shield, signal common, and signal lead. It is correctly called a *common-mode signal*. The signal of interest is called a *normal-mode signal* or *transverse signal*. The second ground is called the *common-mode reference conductor*. A meter or an oscilloscope connected between this reference conductor and any signal conductor will display this common-mode voltage (a voltage common to all leads in a cable).

The shielded conductors in Figure 5.1 carry a signal for conditioning, which conditioning must provide common-mode rejection and perhaps gain, filtering, offset, buffering, or voltage limiting. To be effective, this conditioned signal must be referenced to the second ground. The common-mode voltage must be

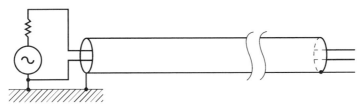

**FIGURE 5.1** Grounded signal source.

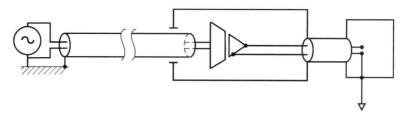

**FIGURE 5.2**  Two shields used in a two-ground system.

rejected if the signal is to have integrity. If there is signal shielding after the common-mode rejection point, it should be connected to the second reference conductor, where the output signal terminates. This use of two shields is shown in Figure 5.2.

There are many circuit techniques for processing signals and rejecting the common-mode content. A common technique uses a differential input circuit. This is a balanced circuit that responds to the normal-mode signal and rejects most of the common-mode signal. The input impedance of the differential stage limits the common-mode current that flows in the input conductors and in the source impedance.

The common-mode rejection problem is reduced significantly reduced if the signal is amplified at the source or if it is a high signal level. If the signal source is a low impedance, any common-mode current flowing in the source impedance may not introduce a significant error. Some environments are very hostile to electronics. In these cases the transducer signals must be brought by shielded cable to a safe point before conditioning can take place.

If an analog signal is converted into digital format before it is placed on a cable, the rejection process changes character. The common-mode level can be a few tenths of a volt and there are no errors. By their very nature, digital signals have fast transition times, which represent a need for bandwidth. Pulses coupled to the line can introduce digital errors. In some facilities, signals are sent digitally over fiber optics paths to completely sidestep the common-mode problem.

When bandwidth is not an issue, the signals can be converted to a current source. A current source has the quality that the current is independent of the common-mode level. A resistor in the current loop provides a voltage that is proportional to the current. This voltage is sensed by electronics using the second ground reference. The signal might be buffered and filtered by the following electronics. This current loop technique can provide a high common-mode rejection. The reason that bandwidth is an issue relates to cable capacitance. Cable capacitance parallels the sensing resistor, forming a time constant. A 1000-ft cable might have a line-to-line capacitance of 0.03 $\mu$F. A 10,000-$\Omega$ sensing resistor limits the bandwidth to about 3 kHz.

In telephony, where voice signals do not require a dc component, transformers can be used to block most of the common-mode signal. Some common-

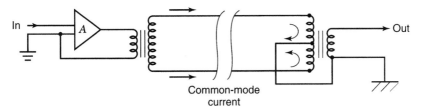

**FIGURE 5.3**   Transformers used on a balanced line to reject common-mode effects.

mode signal can couple across the windings of the transformer through parasitic capacitances. The secondary transformer coil can be grounded at a centertap, providing for a balanced transmission (equal and opposite signals with respect to the centertap). This circuit is shown in Figure 5.3. Any common-mode
current that flows in the same direction in a balanced circuit creates voltages
that cancel at the next transformer. When the telephone lines run parallel to
power conductors, balanced lines provide another benefit. Any coupling to a
power line field induces current to flow equally on both lines of the balanced
circuit and the coupling is effectively canceled.

As discussed earlier, the shield around a cable carrying an analog signal
should be connected to the signal common where the signal grounds. This
shield is often called a *guard shield*, a term applied to a shield used for low-
level or millivolt signals. The guard shield should surround the signal through
any connectors and enter the electronics to protect the input leads. The guard
shield cannot be connected to the second reference ground at low frequencies.
At frequencies above 100 kHz it is practical to connect the shield to the
second ground through a simple *RC* series circuit. This circuit is shown in
Figure 5.4. This connection is a compromise that attenuates the common-mode
signal at frequencies above the bandwidth of the electronics. At these higher
frequencies the shield currents tend to flow on the shield outer surface, due
to the skin effect. At low frequencies the interference currents use the entire
shield and the resulting inner field is the source of coupling. Terminating the
shield at high frequencies is thus a good compromise. Unfortunately, skin

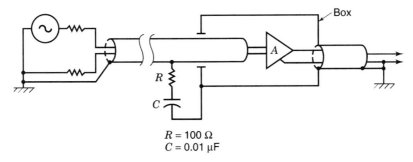

$R = 100\ \Omega$
$C = 0.01\ \mu F$

**FIGURE 5.4**   Common-mode attenuator at high frequencies.

effect phenomena are compromised by braided shield, as the braiding tends to pull surface currents into the center of the cable. The $RC$ shield termination network should ideally be located outside the electronics enclosure. Currents flowing in this circuit are at high frequencies and can radiate into the hardware. Another reason for using this $RC$ filter is to reduce the level of common-mode signal that must be handled by the electronics. In practice, most common-mode rejection circuits have a rejection ratio that falls off proportional to frequency. Having less signal to reject is on an advantage.

## 5.4  COMMON-MODE REJECTION IN INSTRUMENTATION

The processing of many low-level signals is an important aspect of system design and testing. In aircraft structures, missiles, helicopter rotors, and rocket engines, many different types of signals are recorded and/or monitored. These signals must be amplified, filtered, balanced, calibrated, and buffered. Common-mode content must be removed. These signals can be balanced or unbalanced. A balanced signal might come from a strain gage, and an unbalanced signal might come from a thermocouple.

Figure 5.5 shows the signal of interest entering the input shield enclosure. These signals can have bandwidths from dc to 100 kHz. The input electronics could receive its power from a multiply shielded transformer, but this is very expensive. There are circuit techniques that provide operating voltages that follow (are driven by) the common-mode signal. It is also possible to avoid any active circuits in the guard shield enclosure. The signal is connected directly into circuitry associated with the output ground. This is the technique shown in Figure 5.5.

The input impedance to the electronics is represented by values $Z_1$ and $Z_2$. It is interesting to see how high this impedance must be. Consider a 10-mV signal where the desired error level is 0.1% or 10 $\mu$V. If the signal source has an unbalance of 1000 $\Omega$, the limiting current caused by a common-mode voltage would be 10 nA. A common-mode level of 10 V requires that $Z_1$ and $Z_2$ be at least 1000 M$\Omega$. This is a reactance of 2 pF at 60 Hz. This provides some insight into how carefully guarded the input leads must be. If the input leakage capacitance is indeed 2 pF, the reactance at 600 Hz is

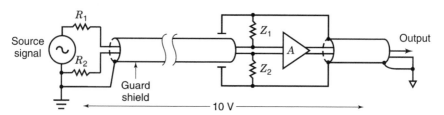

**FIGURE 5.5**  Two-shield system involving a channel of instrumentation.

100 MΩ and the common-mode rejection ratio is reduced by a factor of 10. This example also shows that if a leakage capacitance dominates this reactance, the common-mode rejection ratio falls off linearly with frequency. It interesting to note that 2 pF is the parasitic capacitance across a $\frac{1}{2}$-W carbon resistor.

## 5.5  PROBLEMS

1. An amplifier has a phase shift that is proportional to frequency. The phase shift is 2° at 100 Hz. What is the delay through this amplifier?

2. An amplifier rejects a common-mode signal by a factor of 100,000 referred to the input. The gain is 100. The common-mode signal is 2 V at 400 Hz. What is the output signal?

3. An amplifier rejects a common-mode signal by a factor of 100,000 referred to the output. The gain is 100. The common-mode level is 2 V at 400 Hz. What is the output common-mode signal?

4. The common-mode rejection falls off linearly with frequency. In problem 3, what is the output signal at 4 kHz?

## 5.6  VOLTAGE MEASUREMENT: OSCILLOSCOPES

When a voltmeter or oscilloscope is connected between two points, the voltage that is displayed is often misinterpreted. The display is not the sole result of current flowing in conductors. The signal that is displayed is the result of electromagnetic fields crossing a loop formed by the voltmeter leads. If the loop area is decreased, the resulting voltage is reduced. At dc this is not a problem, as there are no changing fields. To properly measure the voltage drop in a single conductor or along a conductive plane, the metering leads must follow the contours of the conductor.

In intense electromagnetic fields, significant current flows in all nearby conductors, including the shield of the oscilloscope probe. Even if the loop area of the probe is kept under control, the voltage drop in the probe shield will appear as signal. This can be very misleading for an observer trying to understand a problem.

One test to apply is to short the probe tip to the probe shield. If a signal persists, a better probe shield is required. Sometimes a second shield can be added over the existing probe shield. This requires excellent 360° bonding to the oscilloscope common at the panel and at the probe end. In an era of plastic front panels, the connection must be made to an added fitting applied at the panel connector. Another test is to connect the shorted probe to the ground reference of the hardware. If a signal appears, there is excess probe shield current, and again a better probe is required.

The probe shield must be connected to a point near the signal of interest, or another loop can be sensed. Meaningful measurements at high frequencies require good judgment and an awareness of field phenomena. In intense fields the signal of interest can enter through the apertures of the oscilloscope. It is sometimes easier to place the oscilloscope in a separate enclosure than to try to correct this deficiency.

Battery-operated oscilloscopes can provide some relief by eliminating the loop through the power transformer to some remote ground. Unfortunately, the capacitance in free space around a floating oscilloscope still allows probe currents to flow. When the oscilloscope is sitting on a bench, it is grounded through the case capacitance to the bench. Removing the safety ground is usually ineffective, as the path through the transformer is still present. Also note that the power conductor is an inductance and this is in the direction to limit current flow.

The enclosure for an oscilloscope is connected to the power safety conductor. This grounding is required by code. When the oscilloscope is connected to an analog circuit that is already grounded, a *ground loop* occurs, adding greatly to the interference problem. Most users avoid this grounding by using a three-wire to two-wire adapter otherwise known as a power *cheater plug*. When the input probe is grounded to a circuit common, the oscilloscope is again grounded. If the probe common is connected to 300 V, the conductive housing of the oscilloscope is at 300 V and it is unsafe. This is a risk gladly taken by engineers, as it makes their work possible. The same danger exists in a battery-operated oscilloscope. When the probe common is at 300 V, the metal housing is also at 300 V unless it is floating.

The differential inputs of an oscilloscope are not intended to function as an instrument and reject common-mode signals. Oscilloscopes require a common ground connection for the two signals and performs an accurate subtraction A–B or B–A. In an instrument the source ground is a guard shield and it is not connected to a second ground. The oscilloscope can function to reject the average or common-mode signal when the signals originate from the same circuit common. It is not intended to handle signals that originate in another environment. Remember that there can be several common-mode signals operating at the same time. In a strain gage bridge, one-half the excitation voltage is one common-mode signal and the ground potential difference is another.

## 5.7  MICROPHONES

A cable carries the audio signals generated by a microphone to an amplifier. In some cases the microphone is battery operated and the signal is transmitted from the microphone to a receiver. This approach avoids any issue of common-mode coupling. In another design approach the cable carries power to the microphone so that the signal can be amplified at the microphone. This allows high-level signal transmission down the cable over the same conductors.

Microphone signals that are not transmitted or amplified at the source must be carried by the cable as low-level signals. Fortunately, the microphone is not grounded at the source. The person holding the microphone grounds the shield through body capacitance and resistance. Most microphone cable runs are limited to less than 50 ft and the shield currents that flow do not create a problem. It is still preferred to avoid the use of single-conductor shielded microphone cable. Using the shield as a signal conductor can increase the noise coupling.

## 5.8  RESISTORS

There are many types of resistors used in analog design. Stability, accuracy, noise generation, and specific value are parameters that dictate the resistor type that is selected. Carbon resistors are perhaps the most common type used. They are available in standard values from 10 $\Omega$ to 22 M$\Omega$. On special order the range extends from a few ohms to hundreds of megohms. These resistors are available in wattages from $\frac{1}{8}$ to 2 W. Metal film resistors are also available in a wide range of standard values and tolerances. The resistance is varied by changing film thickness and changing the length of a spiral cut into the body of the resistor.

All resistors are noise generators. The random motion of atoms in the resistor generates what is known as *Boltzmann's noise*. This noise increases as the square root of absolute temperature. For resistors below 10,000 $\Omega$, this noise is below 1 $\mu$V in the band dc to 100 kHz. Carbon resistors have a noise figure that is higher than the theoretical value. This noise increases as the resistor is stressed by an electric field. In sensitive circuits where noise contribution is a problem, wirewound resistors are used. Wirewound resistors are made to order to specific values and tolerances. The wire used in construction makes use of special alloys. These alloys form thermocouples at points of copper contact. In input circuits that handle dc signals and where the resistors must dissipate heat or the circuit heats up, these thermocouples can create an unwanted dc voltage. Special alloys can be used to limit this problem. In low-value resistors, part of the coil is reverse wound to limit any inductance. In high-value resistors the wire can be wound in two or more sectors to reduce end-to-end capacitance. These winding techniques are used to provide a resistor that is effective over a wide frequency range.

The geometry of a resistor implies that it has a parallel shunt capacitance. A typical $\frac{1}{2}$-W carbon resistor has a parallel capacitance of about 2 pF. This means that the resistor loses its identity as a pure resistor at high frequencies. A good benchmark to remember is that a 100,000-$\Omega$ resistor can be used in frequency-sensitive circuits up to about 20 kHz. At 200 kHz the resistor limit is about 10,000 $\Omega$. A skillful designer can accommodate this parasitic capacitance and use it as a circuit element and thus create a design using a higher-impedance circuit.

The energy stored in the electric field across a resistor is actually more complex then a simple capacitance might imply. Field lines terminate along the body of the resistor. This is correctly called a *distributed capacitance*. A complete solution requires considering a large group of series and parallel circuits involving portions of the resistance. This is a complex problem that is not worth solving. A single capacitance approximates the performance of most resistors at high frequencies.

In some analog applications, resistors are used as precision attenuators. The exact ratio is important and must be held over a wide range of frequencies. Special metal film resistors for use in this application are available in standard values. These resistors can hold attenuation ratios to an accuracy of 10 parts per million over a wide range of frequencies and voltage levels. The space around these resistors must be protected so that the electric field around the attenuator is uniform. These resistors are usually physically long to reduce the effect of shunt capacitance.

## 5.9  GUARD RINGS

There are cases where current flow must be limited by changes in geometry. On a printed circuit board there are surface currents that result from power supply voltages and signals. These currents are usually very small, but the currents that are being measured are often smaller. One way to limit these currents is to place a conductive ring around a critical point. If this ring is at zero potential, surface currents will not enter the enclosed area. One way to limit any internal current flow is to place a series of grounded *thru holes* along the grounded ring. A ring on both sides of the circuit board will also help. Another technique is to use a contact surrounded by a sapphire insulator. The sapphire has a resistance that is much higher than the printed circuit board materials.

Power supplied through rails to operate railroad trains implies earth current. This current tends to follow buried conductors, which involves building steel. This earth current is concentrated near the point of power generation. Earth current can be quite corrosive (electrolytic action) and techniques are needed to redirect this current flows. In this situation the current flow is three-dimensional.

A ground grid near the point of generation can distribute the current to the surrounding area, thus reducing the flow in the structure. More complex solutions involve monitors that measure the currents in the steel. A bucking current can be forced into the earth to cancel this current. These circuits require feedback to accommodate changes in load and soil conditions. If used, guard rings must be buried in the earth, and they, too, are subject to corrosion.

## 5.10  CAPACITORS

Capacitors were discussed in Sections 2.5 and 3.5. There are literally hundreds of capacitor types. The largest capacitors are perhaps used in the power industry for filtering and limiting reactive current flow. Utilities often supply power to industries that operate many motor loads. These inductive loads require a great deal of reactive energy, energy that is given back to the utility twice per cycle but for which the utility cannot charge the customer. The utility must correct for this reactive current or provide larger power conductors and more generating capacity. The method used to limit reactive current flow is to place capacitors across the power line near or at the user's facility. This is known as *power factor correction*. These capacitors demand reactive current of the sign opposite to the inductive loads. The result is that the utility does not supply a large reactive current from its generators or over its transmission lines. The utility now supplies real power over its lines and industry can operate its motors. Sometimes the utility charges the user for demanding reactive current to offset the cost of supplying correction hardware. During off-peak hours the capacitors must be removed, and this takes additional hardware.

Switching power factor capacitors on and off the line can introduce power transients. One way to limit transient switching current is to connect the capacitors through a series resistor. After the capacitors are connected for a few seconds the resistors are shorted out. This type of correction usually takes place on three-phase transmission distribution systems. Capacitors are switched between phases and ground rather than from phase to phase in a delta configuration.

The smallest capacitors are those inside integrated circuits. The smallest commercial capacitors (components) are perhaps surface-mounted units that can be machine soldered to a printed circuit board. Whatever the size or shape, a capacitor can store field energy. In its simplest form it is two parallel conductive plates separated by a dielectric. The issue in many digital applications is how rapidly this energy can be supplied.

Every capacitor has a series resistance and a series inductance. Mounting large capacitors often involve large conductive loops, and this adds inductance. In digital applications energy may be required in under 1 ns. Some dielectrics are better suited to releasing stored energy in very short periods of time. These decoupling capacitors are often spaced about a circuit board to supply switched energy on demand. The value of a typical decoupling capacitor might be 300 pF.

Capacitors used for rectifier filtering usually use an electrolyte for a dielectric. The electrolyte increases the coupling area between conductors and this increases the capacitance. These capacitors are usually polarized so that there must be an average voltage of one polarity applied across the capacitor. Any power supply ripple voltage across the capacitor implies current flow in the capacitor. Electrolytic capacitors vary from a few microfarads to well

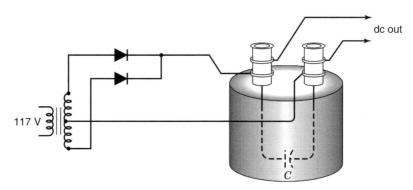

**FIGURE 5.6**   Four-terminal capacitor.

over 10,000 $\mu$F. Electrolytic capacitors are not intended for high-speed applications. At high frequencies the lowest impedance is limited to about 1 $\Omega$. Lower impedances can be achieved by paralleling the electrolytic capacitor with a different class of capacitor.

Electrolytic capacitors often accept charge in short bursts once or twice per power cycle. A full-wave rectifier system may charge a capacitor for only 20% of the power cycle. If the average current taken from the capacitor is 200 mA, the peak charging current might be as high as 2 A. This high value results because the average current for the 20% period must be 1 A. Capacitors can overheat if the charging current is too high.

Current pulses create magnetic fields. Connecting the capacitor correctly can control these fields. The loop area formed by the rectifiers and transformer coil must be kept as small as possible. The capacitor should have four terminals. This means that the connections from the power loop are connected directly to the capacitor. Leads carrying current from the capacitor are placed on the outside of the power connections. The intent is to keep pulse current from flowing in any conductor associated with the filtered voltage. If the capacitor has screw lugs, the geometry is simple. First, the rectified signal is connected to the capacitor. Then the lugs that take the filtered power away from the capacitor are connected. This wiring geometry is shown in Figure 5.6.

## 5.11   PROBLEMS

1. Two 10-$\mu$F capacitors are charged to 10 V and 20 V, respectively. They are arranged in parallel, resulting in a new average voltage. Assume that charge is conserved. Calculate the energy loss. Speculate as to what happened to this energy.

2. A 440-V dc line is switched on to a 10-$\mu$F capacitor through a 5-$\Omega$ resistor. What is the maximum peak current that can occur?

**3.** A factory demands 2 kW of reactive power at 220 V, 60 Hz. A capacitor of what size is required to compensate for this reactive power?

**4.** The ripple waveform across a 100-$\mu$F capacitor is 2 V peak-to-peak at 120 Hz. What is the rms current in the capacitor?

**5.** A filter capacitor does not have four terminals. The shared common impedance on each connection is 20 m$\Omega$. If the peak current is 2 A, what is the voltage that does not get to the load?

## 5.12  FEEDBACK PROCESSES

Feedback is a design technique that uses excess gain to control important parameters. As an example, the gain of a circuit can be made to approximate the ratio of resistors. Excess gain (feedback) can be used to reduce output impedance, raise input impedance, and extend bandwidth. Other benefits include reducing interference and nonlinear effects in active elements. Feedback is not limited to analog circuits. It is the basis of an autopilot, controlling assembly lines, and providing speed control and temperature regulation, to mention a few important areas. Driving a car has many feedback mechanisms at work. Maintaining speed control, lane centering, and car spacing are all feedback processes.

Figure 5.7a shows a typical feedback circuit. This circuit arrangement is called *operational feedback*. The junction of $R_1$ and $R_2$ is called the *summing point*. This point is a virtual ground, as the voltage at this point changes over a very narrow range. If the gain before feedback (open-loop gain) is $-100,000$, the largest signal at the input would be 100 $\mu$V. The gain of this feedback circuit is very close to the ratio $-R_2/R_1$. If $R_1$ is 10,000 $\Omega$ and $R_2$ is 100,000 $\Omega$, the gain is $-10$ (closed-loop gain).

A problem occurs if the gain is to be much higher than $-10$. The parasitic capacitance across a large value for $R_2$ limits the bandwidth. The circuit of Figure 5.7b shows a method of obtaining a gain of $-100$ using smaller resistor values. Assume that $R_1 = 200,000$ $\Omega$, $R_2' = 50,000$ $\Omega$, and $R_3 = 100$ $\Omega$. To find $R_3$, assume that $V_1 = 0.1$ V and $V_0 = 10$ V. The input current $i_{in} = 0.1/20,000 = 5$ $\mu$A. The current in $R_2'$ equals the current in $R_1$. This means that the voltage $V_3 = -i_{in} = R_2' = -5$ $\mu$A $\times 50,000 = -0.25$ V. The current in $R_4 = 0.25/100 = 2.5$ mA. The voltage across $R_3$ is $V_0 - V_3 = 10 - 0.25 = 9.75$ V. The current in $R_d$ is the current in $R_4$ plus the current in $R_2'$. Dividing the voltage by the current yeilds the resistor value $R_3 = 9.75/0.002505 = 3.892$ $\Omega$, so the gain is $-100$. The output attenuator limits the feedback current without raising the value of $R_2$. In this circuit arrangement the parasitic capacitance across $R_2$ does not limit the bandwidth. If high accuracy is required, wirewound resistors would be used.

If the input stage is differential in character, the feedback can be brought to the negative input and the positive input becomes the normal input. The

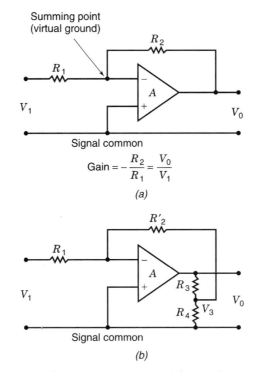

$$\text{Gain} = -\frac{R_2}{R_1} = \frac{V_0}{V_1}$$

*(a)*

*(b)*

**FIGURE 5.7**   Operational feedback.

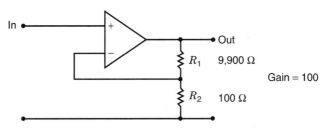

**FIGURE 5.8**   Potentiometric feedback.

input impedance of this circuit depends on the type of input stage, the current required by the input transistor, and the amount of excess gain available for feedback. The input impedance of this type of circuit is generally high. This is called *potentiometric feedback*. If the output attenuator is 100 : 1, the gain of the circuit will be 100. This circuit is shown in Figure 5.8. Assume that the input impedance is 100 k$\Omega$ before feedback. If the total gain available before feedback is 10,000, feedback will raise the input impedance to 10 M$\Omega$. This is a dynamic impedance that does not involve any static input base current.

Feedback can cause instability if the open-loop amplitude response character is not controlled at high frequencies. If the amplitude response falls off linearly with frequency, the phase shift is 90° in this part of the response. If the amplitude response falls off at twice this rate, the added phase shift is 180°. Note that the reversal of sign is already 180°. If the total phase shift reaches 360° and there is still excess gain, the system will oscillate. If the total phase shift is too close to 360°, the step response will have a great deal of overshoot and ringing, indicating that the system is near instability. Most integrated-circuit amplifiers are designed with a controlled phase shift so that there is no sign of instability, even at a closed gain of unity.

## 5.13  PROBLEMS

1. An amplifier has an open-loop gain of 100,000 and a closed-loop gain of 100. If the output impedance before feedback is 100 $\Omega$, what is the output impedance after feedback?

2. The output impedance of a feedback circuit is 0.1 $\Omega$. At 10 kHz the impedance has risen to 5 $\Omega$. What is the apparent output inductance?

3. The input impedance of a potentiometric feedback amplifier is 10 M$\Omega$ at 100 Hz. If the feedback factor is lower by a factor of 10 at 1 kHz, what is the apparent input capacitance?

## 5.14  MILLER EFFECT

The input capacitance to a circuit can be measured by adding a series resistor and noting the effect on frequency response. For example, if the series resistor is 100,000 $\Omega$ and the frequency response drops 30% at 10 kHz, the input capacitance has a reactance of 100,000 $\Omega$ at this frequency. The capacitance equals 15.9 pF.

If a signal from the circuit couples back to the input through a parasitic capacitance, the 30% point will be affected. If the capacitance to a circuit point that has a gain of $-100$ is only 1 pF, the input capacitance will measure 100 pF. It is obvious that a very small capacitance associated with circuit gain can increase input capacitance. This increase in effective capacitance is called the *Miller effect*.

A simple transistor has gain. The voltage gain from the base to the collector can be $-100$. The mutual capacitance from the collector to the base can easily be 1 pF. This means that the input capacitance caused by the Miller effect can be 100 pF. It is good practice in analog circuit layout to give careful attention to these very small capacitances. For example, components attached to the collector and base must have short leads to limit parasitic capacitance.

An extreme example of parasitic coupling exists in a charge amplifier. A charge amplifier uses two capacitors in an operational feedback configuration.

The input capacitor is the transducer and the feedback capacitor determines the gain. Typically, the feedback capacitor might be 100 pF. If the following circuit gain is 100 and the gain accuracy must be held to 1%, the parasitic capacitance to the input is limited to 0.01 pF, a very small value. To achieve this low mutual capacitance the input circuit must be totally enclosed (shielded) in a metal box.

## 5.15   INDUCTORS

Magnetic devices that require turns of wire must have a parallel parasitic capacitance. This parallel capacitance forms a parallel resonant circuit. Manufacturers of circuit inductors specify the natural frequency rather than the capacitance. An inductor with a natural frequency of 1 MHz is probably not accurate above 100 kHz. In many circuit designs the parasitic capacitance can be included as part of the circuit. Above the natural frequency the component looks like a capacitor. To reduce the capacitance the coils of wire can be wound in what are called *pi sections*. Another technique is to weave the conductors into a honeycomb structure to add air space. To limit skin effect, coils can be wound using *Litz wire*. Litz is a conductor composed of multiple strands of very fine insulated wire. This wire reduces the series resistance of the inductor, making it closer to an ideal element. (See the discussion of the skin effect in Section 4.12.)

Inductors that store energy in an intentional air gap require a high-permeability magnetic path to focus the field. At high frequencies where the permeability falls off, the inductance may change. Care must be taken when inductors carry large currents at dc or at power frequencies. An inductor can handle only a limited number of ampere-turns before the core saturates. For this reason inductors used in high current applications often have no core. A few turns of wire supported by the conductor itself can be an effective inductor. Without a core the inductor cannot be saturated. Without a magnetic core an inductor can be a rather bulky component.

Ferrite beads are sometimes used at high frequencies to form small series inductors. The technique is to thread one or more beads on a signal conductor. Sometimes the conductor threads one bead with two or three turns. These small inductors can add only a very small series impedance and can be effective only in low-impedance circuits. Sometimes the spacing afforded by the beads limits capacitive coupling, and the same effect can be produced by using a plastic bead.

A multiconductor cable can be wrapped as turns on a large ferrite core. This geometry forms a series inductance for limiting common-mode currents in the cable (currents flowing in the same direction in all conductors). This geometry offers no significant impedance to the normal-mode signals handled by the cable. The number of turns is usually limited and the core must be fairly large to be of any practical use. A few turns adds a small inductance, which again means that the technique can be effective only in low-impedance circuits. It is preferable to allow common-mode current to flow on the outside surface of a proper shield. If this shield is correctly bonded at the termination, common-mode currents will not enter the hardware. This further implies that there is a significant radiation problem in the area or these unwanted currents would not be present.

An inductor with two coils (four connections) can be used to limit common-mode current flow at high frequencies in signal conductors. The two coils are placed in series with the two signal conductors. This component is called a *balanced inductor*. The coil geometry is such that there is a reactance added to common-mode currents. Normal-mode signals are not affected. This technique requires a shunting path for common-mode current after the balanced inductor. Without this shunting path the balanced inductor is ineffective.

## 5.16   TRANSFORMERS

Basic transformer action was discussed in Section 2.18. The wire that forms the coils has capacitances between adjacent turns. Smaller capacitances exist between nonadjacent turns as well as between layers of wire. The problem is further complicated by mutual capacitances between the individual coils of wire in separate windings. The voltage impressed across the primary coil must supply reactive energy to all of these distributed capacitances. The leakage capacitances between windings allows for common-mode current to cross the transformer. The effect of these distributed capacitances is often simplified down to several capacitances: one shunting the primary coil, a second shunting the secondary coil, and a third showing a reactive path between the coils. In most practical situations this representation is adequate.

The magnetic flux that couples windings of a transformer is determined by the geometry. In transformers with a magnetic core, the $B$ flux is concentrated in the core. There is always primary flux that does not couple to turns of the secondary. For example, the field that surrounds each turn does not couple to the secondary coil. Also, there is always some field that avoids the core completely. This noncoupling field is referred to as *leakage inductance*. Leakage inductance is often drawn as a circuit element in series with any coil resistance. This inductance exists primarily in air and is a linear element.

The coils of a transformer are usually wound on the same bobbin. This geometry limits leakage inductance. The primary coil is usually wound on the bobbin first, which places it closest to the core. Placing the primary and secondary coils on separate bobbins increases the leakage inductance. This separation technique is sometimes used to reduce common-mode coupling and avoid the use of shields. The coils both have capacitances to the core, but when the core is grounded the bridging or mutual capacitance can be made very small. This construction can be used to isolate small amounts of power without the use of shields. Separate bobbins also increase the voltage rating between coils.

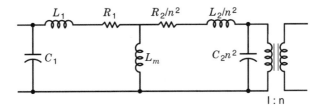

**FIGURE 5.9**   Equivalent circuit of a transformer.

The equivalent circuit for a transformer is shown in Figure 5.9. The transformer at the right of the diagram is ideal. It has no leakage inductance, requires no magnetizing current, and has no coil resistance. It serves to isolate the voltages and provide a turns ratio. Any circuit element can be referenced to the other side of the ideal transformer by multiplying the impedance by the square of the turns ratio. If the transformer has a step-up ratio of $1:2$, a series resistance of 2 $\Omega$ on the primary side is equivalent to 4 $\Omega$ on the secondary side. A load resistor of 100 $\Omega$ on the secondary side is equivalent to 25 $\Omega$ on the primary side. A shunt capacitance of 1000 pF on the primary side is equivalent to 250 pF on the secondary side. It is obvious that a large turns ratio reflects a big change in impedance. A $10:1$ step-up ratio makes a $0.01$-$\mu$F shunt capacitor on the secondary appear as a $1.0$-$\mu$F shunt capacitor on the primary.

Leakage inductance can create a serious field problem for large facility distribution transformers. Secondary load currents can often be as high as 500 A, and this current flows in the leakage inductance. This inductance implies a field that is external to the transformer core. These fields can be extensive enough to affect the performance of computer monitors in the area. Shielding against these fields (near induction fields) can be very difficult and expensive. Transformers that must be placed near sensitive circuits must be specified as to their leakage inductance field. The flux level inside the core depends on the core material. The flux level outside the core depends on the leakage inductance on the transformer loading and on winding geometries. A larger core and fewer turns are in the direction to reduce the leakage field.

A second problem that exists for all large power transformers involves fields that couple into nearby structural conductors. Power transformers should never be mounted by the core such that the mounting hardware forms a shorted turn around any part of the core. If the core is used for mounting, some sort of insulator must be used to break up the shorted turn. If the transformer is mounted on or near building steel, a conductive loop formed by the steel can couple to leakage flux from the transformer. This can result in power currents flowing throughout a building. The leakage flux in electronic facilities is usually rich in harmonic content, as the loads are usually nonlinear. This means that the current flowing in building steel will also be rich in these same harmonics. It is preferable to mount these transformers away from building steel on some sort of wooden structure.

The equivalent circuit for a transformer indicates how to measure leakage inductance and magnetizing inductance. The leakage inductance is measured by shorting all secondary coils together. The voltage–current relationship on the primary defines the leakage inductance. (Use an external signal generator to make the measurement.) The frequency should be high enough so that the series resistance is smaller than the reactance of the inductance. Magnetizing inductance can be measured with all the coils open circuited. The current that flows when the power is connected defines the magnetizing impedance. (This measure can be made by using the utility power.) In general, the magnetizing current will not be sinusoidal and the term *impedance* is somewhat incorrect. A reasonable figure can be made by noting the ratio of peak voltage to peak current. In general, this ratio will change with voltage level. This technique also measures permeability. The voltage defines the *B* field and the current defines the *H* field. Knowing the number of turns and the core dimensions, the permeability can be calculated.

The equivalent circuit for a transformer does not show the leakage capacitance that exists between the primary and secondary coils. This equivalent circuit is only intended to represent the transformer at power frequencies. A single leakage capacitance is only an approximation. Even this capacitance should have a series voltage determined by coil geometry.

Switching regulators often operate using square waves with repetition frequencies above 50 kHz. Transformers in this frequency range require very few turns. Reducing turns is in the direction to limit leakage inductance. If the core material surrounds the coils, the external field is further controlled. It is preferable to design the transformer to operate at maximum flux levels to limit the number of turns involved. The leakage flux transitions in microseconds and can easily couple into nearby circuits. To control fields further, all connections to the transformer should have limited loop areas. The number of turns in a transformer winding might be 10 and an external loop having the same loop area as one turn could create a significant field external to the transformer. The wiring geometry used to connect rectifiers and filter capacitors to the transformer is critical at these high frequencies. Fortunately, the capacitors can be small in size and value, as the filtered energy does not need to be stored for more than one cycle.

Shields are not practical in these transformers. The rapid voltage transitions demand too much reactive current. Coupling interference can be reduced only by the careful balancing of winding geometry. Current pulses can be canceled through symmetry.

## 5.17 PROBLEMS

1. A switching regulator supplies current to a capacitor 50% of the time. The clock rate is 50 kHz. If the voltage is 10 V and the load is 100 $\Omega$, how

large must the filter capacitor be to keep the voltage drop to 0.1 V? Assume square waves.

2. The rise time on a switching regulator voltage is 0.2 $\mu$s. The square-wave voltage is 20 V peak. The loop area to the capacitor is 3 cm$^2$. The average current drawn from the capacitor is 100 mA. The capacitor is charged 50% of the time. What is the radiation level at 3 m?

3. A transformer has one step-up turns ratio of 1 : 3 and a second step-up turns ratio of 1 : 2. If the loads are 200 and 400 $\Omega$, respectively, what is the primary reflected impedance? What power is dissipated if the primary voltage is 12 V? If the two coils have 100 pF of capacitance, what is the reflected primary capacitance?

## 5.18   ISOLATION TRANSFORMERS

The word *isolation* can be applied to many areas. There is isolation from ground, from electric or magnetic fields, from ground motion, from wind, from noise, and so on. The question is: What kind of isolation does an isolation transformer provide? Isolation transformers are often added to racks of hardware because it is felt that the power source is a source of interference. The word *isolation* is intriguing, and so the transformers are installed. This addition usually does little to improve performance.

Isolation transformers had their start as power transformers in the early days of instrumentation. These transformers had three shields: a primary shield, a guard shield, and a secondary shield. These shields were necessary to limit parasitic power currents and common-mode currents from flowing in input signal circuits. There was one transformer required for each instrument. The idea of shields doing something valuable caught on with transformer makers, and the term *isolation* did wonders to improve sales. With the word *isolation* the assumption was made that the transformers would isolate the user from his or her problems. These transformers are available commercially with multiple shields brought out for users to define.

An isolation transformer can be used to power a group of instruments or a rack of electronics. The main benefit is to limit the flow of common-mode current. This requires that a middle shield be connected to the local hardware ground. The primary shield should be connected to the grounded power conductor. A third shield (associated with the secondary winding), if it exists, should be connected to one side of the secondary. All three shields can be connected inside the transformer housing. This is not a guard shield approach to isolation. Guard shield applications require one transformer per instrument.

Separately derived isolation transformers are discussed in Section 7.8. These transformers are "listed," and as facility transformers the secondary coil must be grounded. (There are special cases where grounding can be avoided.) This grounding must conform to code. The secondary of an isolation transformer that is added inside a rack of hardware is often not grounded. This meets code if it is inside user hardware. This is not ideal because a fault to this secondary will not trip a breaker, and the result could be a shock hazard.

The primary and secondary connections should never be made in the same junction box or panel. This adds to the mutual capacitance between primary and secondary coils and negates the value of the isolation transformer. The conduit and housings (equipment ground) are all connected together, and this is acceptable.

## 5.19 SOLENOIDS AND RELAYS

Solenoids are designed to perform mechanical work when energized. A moving arm closes a magnetic gap when the solenoid is energized. This motion reduces the energy stored in a magnetic circuit. Solenoids can be used to trip latches, operate valves, or operate electrical switches.

Relays are specifically designed to open or close one or more electrical contacts. There are latching relays that assume a mechanical state until unlatched. These relays are used in circuits where sustained power may not be available. Relay contacts in low-level circuits form thermocouples. Heat flowing out of the operating coil can raise the temperature of the contacts, introducing dc errors. A latching relay avoids this heating.

All relays and solenoids coils are inductors. The energy stored in this inductance cannot be disregarded when the relay or solenoid is turned off. The stored energy requires the current to continue flowing. At the moment of disconnect the current starts to decrease. The voltage across the coil (inductor) is proportional to this rate of change in current. Because the current is decreasing, this voltage is of opposite sign to the operate voltage. Because the decrease is rapid, the reverse voltage can be thousands of volts. This large voltage causes arcing across the operating contact, and this is a source of interference.

When the operating contact opens up the coil does not operate as a simple inductor. It is an inductor in parallel with a parasitic capacitance. This is a parallel resonant circuit. At the moment the contact opens the current continues to flow but in the parasitic capacitance. The voltage increases in the capacitance until the energy that was stored in the inductance is all transferred to the capacitance. This takes approximately one-fourth of a cycle at the resonant frequency of the coil. An example can illustrate the voltage rise across the inductor. Assume an inductance of 1 H and a parasitic capacitance of 100 pF. If the current in the relay is 100 mA, the energy stored in the inductance is $\frac{1}{2}LI^2$. This is 0.005 J. The resonant frequency is 15.9 kHz. One cycle takes 62.8 $\mu$s. One-fourth of a cycle takes 15.7 $\mu$s. If fully transferred to the capacitance, this energy results in 10,000 V. There is a race between this rising voltage and the opening of the contact. If the voltage wins, there will be an arc across the contact. In this example the voltage rise time is so short that the

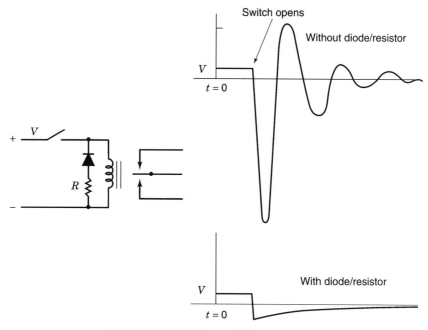

**FIGURE 5.10**  Relay and damping diode.

contact will probably arc. The energy stored in the inductance will mainly be lost in heat in the arc. The arcing process is a pulse of current that can radiate over a wide spectrum.

There are two ways to limit the high voltage after contact opening. A capacitor placed across the coil will lower the natural frequency and the peak voltage. An added capacitance demands a high current when the relay operating voltage is applied. This is a drawback to this solution. A second method is to place a reverse diode across the coil. When the voltage reverses across the inductance, the diode shorts out the coil. The current must then decay in the internal resistance of the relay. This $L/R$ ratio (time constant) can be large and the relay may take a long time to release. A compromise exists if a resistor is placed in series with the diode. This circuit is shown in Figure 5.10.

In power circuits where switching involves inductive loads, the switch is often spring loaded to open quickly. If the contact is in oil, the $E$ field is locally reduced and the resulting reverse voltage can be tolerated. Arcing causes pitting of contacts and it should be avoided if possible.

## 5.20  PROBLEMS

1. The inductance of a 24-V relay is 0.5 H. The parasitic capacitance across the inductance is 400 pF. The resistance $R$ of the coil is 1000 $\Omega$. What is

the maximum voltage across the coil after the switch opens? If the switch arcs across, how much energy must be dissipated in the arc? Assume no $R$ losses.

2. In problem 1, assume that the value of $R$ is 100 $\Omega$.

3. In problem 1, if a diode clamps the relay coil upon the switch opening, what is the time constant for current decay in the relay? Assume that the relay drops out in two time constants. How long will the relay remain operational?

4. In problem 3, assume that $R$ is 100 $\Omega$.

## 5.21   POWER LINE FILTERS

Power line filters are a part of many hardware designs. The term *filtering* implies that something is removed from the power line. The hope is that interference brought to the hardware will not pass through the filter. To some extent this idea is valid, but there is more to be considered.

A simple filter consists of series inductors and shunt capacitors. The inductors oppose the flow of rapidly changing current and the capacitors shunt aside any rapidly changing voltage. In a single-phase circuit the "hot" conductor can be filtered with respect to the grounded conductor. This means that the hot conductor has a series inductor and the capacitor that follows connects to ground (the grounded conductor). This circuit is shown in Figure 5.11. Viewed as an $LC$ circuit the amplitude response above a certain frequency falls off as the square of the frequency. This means that interference that is superposed on the power line is attenuated. Note that the inductor and capacitor form a series resonant frequency. The load has the effect of limiting energy storage in the capacitor so that resonance phenomena (peaking) will not occur.

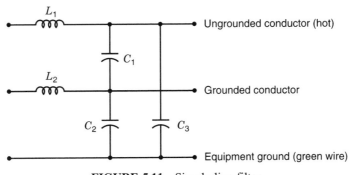

**FIGURE 5.11**   Simple line filter.

The series inductor has a parallel parasitic capacitance. This limits the attenuation of the filter at some upper frequency. To obtain further attenuation a second inductor and capacitor can be added. If this inductor is smaller in value, its natural frequency will be much higher. This allows further attenuation at high frequencies. Many commercial line filters often have just one series inductor. This is done to reduce both cost and size. A small inductor means that the filter can only be effective at higher frequencies. Typically, attenuation is provided for interference above 100 kHz.

In some filters, capacitors connect between the power conductors as well as from line to housing (equipment ground). The line-to-housing connection attenuates common-mode signals. If the equipment ground is not brought to the equipment or if the equipment ground conductor is a long separate conductor (inductive), the common-mode filtering may not be effective (see Section 7.4). It is illegal to place any component in series with the equipment ground.

## 5.22  REQUEST FOR ENERGY

When a light switch is activated, the energy required by the light must ultimately come from a generator. At the moment of switch closure the voltage across the switch implies an $E$ field. The power conductors represent a charged transmission line where the termination has been changed in a step manner. Two waves are propagated from the switch, one toward the light and another toward the power source. As the wave travels toward the generator, it picks up field energy at every parallel branch along the way. There are reflections at every junction and load point. Each reflection sends a wave in two directions. These waves go into other power loads and back toward the light. These many reflections tend to smooth out the request that eventually must propagate back to the generator. The generator fills in the needed power and the light burns on.

At the switch, the voltage momentarily falls to half-value. This is what happens when a charged 100-$\Omega$ transmission line is connected to an uncharged 100-$\Omega$ transmission line. This 50% voltage drop can actually be measured near the switch. At a nearby breaker this waveform is smeared and reduced in amplitude. At the service entrance the transient is hardly noticeable, and at the local distribution transformer it is represented by a slight sag in voltage. In effect, the energy is pulled from the local electric field. The series of waves that propagate backward involve both fields, as there are moving charges. These waves arrive at the generator at different times as a result of many reflections. This accounts for the smearing of the initial request. The generator adjusts to supply the added field energy required to operate the light. No new energy can arrive on the power grid until a generator supplies it. A 10-mile round trip at half the speed of light take 0.1 ms. It all appears instantaneous, but that is not the case. All of this happens in less than one electrical degree at 60 Hz.

## 5.23 FILTER AND ENERGY REQUESTS

One role of power line filtering is to supply step demands for energy from a local source. This stops the initial request from propagating a steep wavefront into nearby hardware. A filter with a terminating capacitor can supply this immediate energy. This type of filter limits the nature of the power request that goes back toward the generator. A capacitor as a source of energy is not the same as a low-impedance source over a wide bandwidth. The utility in some cases can appear to have a source impedance below $0.1\ \Omega$ from dc to 100 kHz. A $1\text{-}\mu\text{F}$ capacitor has a reactance of $1.59\ \Omega$ at 100 kHz. At 10 kHz the reactance is $15.9\ \Omega$. To make the filter function, the series inductance can not raise the line impedance significantly at 10 kHz. An inductor of $15.9\ \mu\text{H}$ has a reactance of $1\ \Omega$ at 10 kHz. This example indicates that a filter designed to limit load-to-line interference may be different than a filter designed to limit line-to-load interference.

## 5.24 POWER LINE FILTERS ABOVE 1 MHZ

In hardware it is often necessary to limit the penetration of electromagnetic energy. The power lines are one source of this energy. The energy (interference) can use any conductor pair for transmission. This includes hot leads, grounded leads, neutrals, and green wires. Frequently, energy is transported between power conductors and ground (earth). The energy can arrive in differential or common-mode form. In common-mode interference the transmission line consists of power conductors and a reference ground. The reference ground is usually the hardware cabinet.

A mistake that is often made is to run power conductors to a filter located inside the hardware. Any conductor that carries interference current inside the hardware will radiate energy into the hardware. This bypasses the filter completely at high frequencies. The filter can be inside the box mechanically, but it must be outside the box electrically. All interference fields must cause current to flow on the outside surfaces of the hardware.

The filtering of common-mode interference requires attention to each lead. An inductor can be placed in series with each power lead, including the neutral. An inductor cannot be placed in series with the safety or green wire. This violates the National Electrical Code. This means that the green wire must be terminated outside the hardware and continued inside the box from the other side of the barrier. This geometry is shown in Figure 5.12.

## 5.25 MOUNTING THE FILTER

The performance specifications of filters often extend above 100 MHz. The testing of a filter provides some insight into how the filter should be used. A

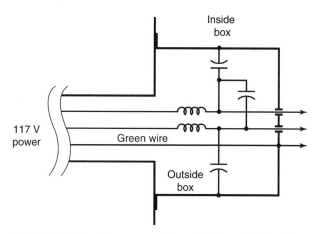

**FIGURE 5.12**   Common-mode and differential filtering.

filter, signal generator, and meter are all bonded to a ground plane. The leads to the filter are shielded with the shield bonded to the ground plane. There are no openings in this input shield. The only fields that leave the filter are between the output power conductors and between these conductors and the ground plane. The green wire is bonded to the ground plane outside the filter.

Several tests can be made: the attenuation of normal-mode signals, the attenuation of common-mode signals, and the conversion of normal-mode signals to common-mode signals. If the filter is not well bonded to the ground plane, current in this connection can cause a common-mode signal to be measured on the output of the filter.

The fields associated with the power leads are maintained inside the input shield and inside the filter. The filter components store a great deal of field energy, but some of this energy fills the space between the components. The output leads can couple to some of this field, thus bypassing the filter. A good filter design involves inner filter partitions and careful attention to every geometric detail.

If a filter is to be effective in its application and function over its specified frequency range, the following steps need to be taken:

1. Do not bring the equipment ground (safety wire) inside the hardware. Terminate it externally.
2. The power conductors must be shielded (inside a cable) by the enclosure potential if brought inside the hardware.
3. The filter should limit field entry on all power conductors (except the safety wire).
4. The filter housing must be properly bonded to the hardware enclosure. If the filter is mounted inside the hardware, the bond to the hardware becomes more critical. Filter currents cannot penetrate a metal plate.

This means that surface currents must flow through the lead exit hole to get to the hardware surface. If the filter mounting forms an aperture, fields from the filter can still enter the hardware. To avoid this problem, the exit hole should be bonded to the hardware.

## 5.26 OPTICAL ISOLATORS

Optical isolators are often used on control lines. This type of logic coupling solves the common-mode problem. These isolators should not be placed inside the hardware or the leads can bring interference into the hardware. Adding line filters raises the cost. The isolator elements should be mounted so that they straddle the hardware interface. This requires special attention during design.

## 5.27 HALL EFFECT

The *B* field exerts a force on a moving charge. This force can be used to divert current flowing through a semiconductor. The diverted current that is collected can be used to measure the intensity of the magnetic field. This diversion of current is called the *Hall effect*. The current that is collected is proportional to the *B* field and to a voltage. This makes it possible to use a Hall effect device as an analog multiplier.

## 5.28 SURFACE EFFECTS

Skin effect dominates the performance of many circuits at high frequencies. A good example is the performance of high-frequency power transistors. Manufacturers often provide exact geometries for the layout of oscillators or power drivers. They will even specify the size and location of decoupling capacitors. In some cases the component housing must make continuous contact with the ground plane. If the component is riveted to the surface without a proper gasket, the circuit will malfunction. The current cannot flow smoothly to the component and the added impedance (field generation) modifies the performance.

## 5.29 REVIEW

All electronic components are associated with parasitic processes that influence their high-frequency performance. Resistors and inductors have shunt capacitances. Transformers have leakage inductance, magnetizing inductance, shunt capacitances, and leakage capacitance across the coils. Parasitic effects allow common-mode currents to flow. Common-mode rejection plays a big part in selecting an approach to handle all signals. The application of filters to a power line illustrates how difficult it is to keep interference from entering hardware.

The measurement of signals using an oscilloscope can prove difficult. An understanding of field phenomena is needed to interpret the observations. Sometimes the patterns observed include fields in the environment. This is the result of loop coupling and probe-shield current coupling.

The processing of digital signals aside from their logic content is analog in nature. Rise time, delay, interference coupling, and filtering are all analog issues. Power supplies and their associated transformers, rectifiers, and filters involve analog processes. These circuits should be constructed to avoid radiating any interference.

# 6 Digital Design and Mixed Analog/Digital Design

## 6.1 INTRODUCTION

In this chapter we discuss the design of digital hardware exclusive of integrated circuit design. Topics such as board design, cabling, connectors, housing, ground planes, and decoupling capacitors are considered. Design needs keep changing, but the basics remain. The demands of mass production result in the development of large-scale integrated circuits. Some of the problems encountered when the circuits were made from discrete components are reduced. One benefit is that multilayer boards are not always required.

Designs that require the use of many components still require attention to detail. As circuit speeds increase, the problems of radiation become more significant. Interconnecting many peripherals or computers can prove to be a challenge. These problems are all addressed in this chapter.

## 6.2 LOGIC AND TRANSMISSION LINES

When a logic switch changes state, a new voltage is sent to one or more logic inputs. The transport of this voltage is via one or more short transmission lines. To understand the sequence of events, it helps to consider a very simple case. In Figure 6.1 the power supply for the logic switch is located at a distance from the logic. The logic signal must be sent forward to another logic element. The connecting conductor (trace) and the return conductor (ground plane) represent an output transmission line. Similarly, the connection from the power supply is via a trace and a return conductor (ground plane) and this is an input transmission line. When the switch changes state, energy is sent forward. The only energy available comes from the stored electric field in the input transmission line. If the characteristic impedance of the two lines is both 50 $\Omega$, then the voltage must drop to half. The energy moving forward comes from the electric field in the input line. When the reverse traveling wave reaches a power supply capacitor, there is a reflection and the correct voltage is sent forward. Meanwhile, the voltage sent to the next logic element is one-half of its correct value. It obviously takes time before the correct voltage can appear and the reflections subside. If there is more than one transmission line

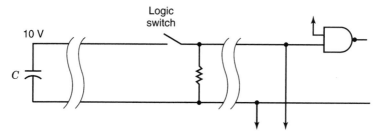

**FIGURE 6.1**   Simple logic switch.

(a fan out), there are more reflections and it takes even longer to establish the correct voltage.

There are two problems associated with this circuit geometry. First, the forward traveling logic level is only half value. Second, the supply voltage has dropped to one-half value and this might affect the operation of other logic elements using the same transmission line. It is easy to see how complex this situation can get if many logic elements operate at the same time. The power supply line can drop to near zero if 10 logic lines try to drive transmission lines at the same time.

Clocked logic circumvents many of the problems associated with these transmission line delays. At a clock rate of 10 MHz, there is 100 ns for the many reflections to run their course. After this delay the next logic state is validated by the clock signal. If the clock rate is raised, there may not be enough time for the reflections to settle out and the logic will not be functional.

The solution that is universally adapted is to supply points of local energy storage near the logic elements. Small capacitors across the power supply lines near the logic supply energy for forward transmission. When a logic switch closes, the energy comes from a local capacitor, not the incoming power transmission line. These capacitors are called *decoupling capacitors*.

## 6.3   DECOUPLING CAPACITORS

The value of decoupling capacitors can be determined from the energy requirements. On a typical logic line the capacitance is around 1 pF. If 10 logic lines operate at once and the voltage is to sag less than 10%, the decoupling capacitor need be only 100 pF. A factor of 3 provides a margin of safety and allows for line loading. Typical values might be 300 pF. The number of capacitors is determined by logic type, the number of logic gates that fire, and clock rates. In an integrated circuit hundreds of gates might transition at the same time, and this requires much more local energy. The manufacturer of the integrated circuit will often specify the degree of decoupling that is required.

The lead length of a decoupling capacitor determines its parasitic inductance. Smaller components have less inductance and are preferred. The best capacitors are surface mounted without leads. Dielectrics vary in the time it takes to release stored field energy. Decoupling capacitors should be selected based on this release time.

The power supply to a circuit board should be decoupled by at least one large capacitor. The energy supplied to the board should flow to the board on a leisurely basis so that this transmission is not an area of radiation. In troubleshooting it is always wise to look at the power supply leads and make sure that they do not carry voltages with fast transitions.

Decoupling capacitors came in many forms. Aside from individual capacitors, there are single-in-line packages (sips) that house a group of capacitors. There are arrays of capacitors in a standard integrated circuit format. Some of these units solder in place under a standard integrated circuit and thus use up less board space. These capacitor configurations should be evaluated before they are used.

## 6.4  GROUND PLANES

The logic traces in Section 6.3 were placed over a ground plane. This plane provided a return path for every logic signal. In most layouts it is impossible to work with a two-layer board consisting of interconnections and a ground plane. The crossing of traces requires many external jumper wires, and this is impractical on a single surface. The use of a ground plane has become a necessity with higher and higher clock rates. This has meant the addition of a third and fourth circuit layer. The fourth layer can be a power plane (it is also a ground plane for fields) supplying power to all the circuitry.

Theoretically, the only areas used for current flow on the ground plane are directly under each logic trace. There is no advantage to providing individual return traces. If there is signal cross-coupling, it is the result of fields sharing the same physical space. The ground plane approach works because the fields under each trace are well confined. Remember that current flow is limited to areas where there is field concentration.

Larger component counts result in a higher trace count. To keep the component density high, additional layers are needed to handle the traces. It is not uncommon for the layer counts to exceed six. This poses a problem for the designer. Designs cannot be made functional unless they are close to their final geometry. This forces the designer to commit to a final package before hardware testing can occur. If there are mistakes, the design costs can be excessive. This implies that designs must be tested in software before there is a commitment to hardware. Logic software cannot test for radiation and susceptibility.

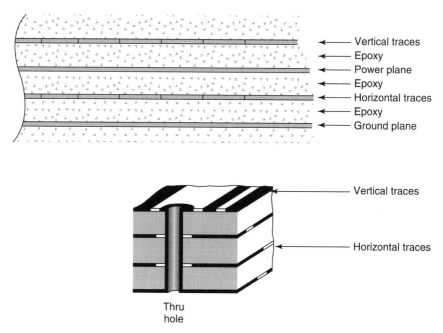

**FIGURE 6.2**   Use of a ground plane and a power plane.

## 6.5   POWER PLANES

The use of a power plane to bring power to the components has several advantages. The power plane functions as a ground plane with an added static field. The ground plane and power plane form the two plates of a capacitor. A layer of traces can be placed between the planes. The fields from these traces are tightly confined and do not radiate, as there is no field above or below the board. If there is radiation from these traces, it is out the edge of the board. One example of a ground plane/power plane geometry is shown in Figure 6.2. Another approach might be to place the two planes in the center of the stack. Designers prefer to limit the power plane to an inner surface.

The advantages of a power plane are:

1. Radiation from internal traces is greatly reduced.
2. The parallel planes provide some fast decoupling capacitance.
3. Fields from these buried traces cannot couple to traces on the other side of the ground planes.

## 6.6   DECOUPLING POWER GEOMETRIES

It is possible to supply power over a very low impedance transmission line. If the ground plane/power plane is a 4-in.-wide strip with a spacing of 0.005 in.

wide and the dielectric constant is about 6, the characteristic impedance will be below 0.1 $\Omega$. Energy can be drawn from this line over a wide frequency range without the voltage sagging. If this line terminates properly on the ground plane/power plane for the circuit board, there may be no need for decoupling capacitors.

Another solution to decoupling a circuit board is to use added power planes and ground planes just to store local energy. This approach avoids the need for decoupling capacitors and their associated thru holes, thus reducing the board area requirements. This type of board design is commercially available from some manufacturers.

## 6.7  GROUND PLANE ISLANDS

Two-sided boards and digital circuitry are not often compatible. In an effort to provide a ground plane and use the same surface for trace crossings, the ground plane is often broken up into islands. These islands of ground are interconnected by traces. The problem with this approach is that it creates large loops for some of the logic paths. These loops are radiators and they can also couple to external fields. The ground plane is not an electrostatic shield in the analog sense. It is the return path for all transmission lines. Compromising these paths can lead to signal delays, radiation, and susceptibility problems.

## 6.8  RADIATION FROM LOOPS

There are three loop configurations that can allow energy to radiate from a logic circuit board. The first source involves the circuit trace loops and the ground plane return. The second source involves the integrated circuit (IC) die (chip) and its circuit connections. The IC die rests above the board in the center of a header. The loop in question involves the legs of the IC and the distance to the die. The loop area includes the power leg, the output signal leg, and the ground plane. This geometry is shown in Figure 6.3. The third loop involves the decoupling capacitors. If the decoupling capacitors are surface mounted, this area is minimum.

## 6.9  PROBLEMS

1. Assume that a circuit board requires ten 500-pF decoupling capacitors. The board is $10 \times 20$ cm. If the spacing between a ground plane and a power plane is 0.01 in., what dielectric constant is required to provide decoupling in the ground plane/power plane geometry?

2. A 100-$\Omega$ unterminated transmission line is connected to a 10-V source 10 ft away. A 10-$\mu$F capacitor is connected to the open end of the line.

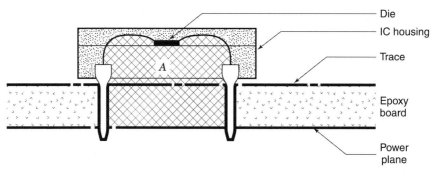

Note: The radiating area is $A$

**FIGURE 6.3**   Radiating loop on a circuit board.

What happens in the time it takes the wave to propagate to the source and return?

**3.** A circuit card draws 20 mA at 10 V. How large must the board decoupling capacitors be to supply this energy for 1 ms and maintain the voltage to within 90%? Assume that the current demand is constant.

## 6.10   LEAVING THE BOARD

The digital lines that leave a circuit board are no longer associated with the ground plane of that board. Conductors carrying logic have fields that must be associated with some other return path or ground. If the signal lines are a part of a ribbon cable, the ground return can be one or more conductors in the ribbon cable. This geometry has added loop area, so this can be a source of radiation.

Ribbon cables come in a variety of geometries. Conductors can be carried next to a ground plane located on one side or on both sides of the ribbon cable. The conductors can be individually shielded with the shields isolated or bonded together along the run. The most common ribbon cable is a group of parallel conductors with no shields provided. If a ground plane is carried on the side of the ribbon cable, it must be bonded to the circuit board to be effective. One way to effect a good bond is to use an entire row of pins on the ribbon cable connector. The pins must all connect to the ground plane. This adds to the pin count and to the size and cost of the connector. The same practice must be followed on both ends of the ribbon cable. Without this treatment, the shielded ribbon cable will be ineffective. The shield is not intended to be electrostatic in nature. It is intended to provide a near return path for the transmission of high-speed signals.

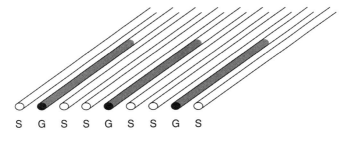

S,Signal; G,Ground return

**FIGURE 6.4**   Assignments of ground returns on a ribbon cable.

The most effective way to utilize the unshielded ribbon cable and minimize signal loop areas is to provide a ground trace next to each signal. This practice is shown in Figure 6.4. The majority of the return current will take the near path as this stores the least field energy. Some will argue that a parallel set of ground connections creates ground loops. That is true. A ground plane is an infinite number of ground loops, and that is why it works so well. Ribbon cable is not intended for use with low-level analog signals. Ribbon cables are a very effective way to transport digital signals. Ground loops that allow undefined currents to flow in analog signal conductors should still be avoided.

## 6.11   RIBBON CABLE AND COMMON-MODE COUPLING

When a ribbon cable leaves a circuit board, it is often many inches away from the equipment housing. If there are electromagnetic fields in this space, common-mode signals are coupled to the cable/housing loop. Common-mode currents flow in the ribbon cable conductors, as there is no protective shield to carry the current. The level of interference depends on loop area, field intensity, and the sensitivity of the input circuits.

   To limit common-mode coupling, the ribbon cable should be routed against the enclosure ground plane. This reduces the common-mode coupling loop area. Excess cable should not be wrapped in a roll, as this adds to the loop area (see Section 4.29). In applications where bandwidth can be compromised, small passive filters can be used to attenuate common-mode content. A logic request from a switch closure or a keyboard stroke can obviously be delayed 1 ms without causing a problem.

## 6.12   BRAIDED CABLE SHIELDS

When signal conductors are carried in a shielded cable, the quality of the shield can be an important consideration. Cables intended for analog applications are often shielded by a conductive braid. This allows flexibility and allows for shield termination (soldering). The woven braid consists of fine strands of tinned copper wire. The tightness of the braid is one consideration. For a thin weave, the leakage capacitance is higher and this allows for more interfering field coupling.

Braid has the advantage of flexibility but the disadvantage of allowing currents to penetrate the shield. At low frequencies, currents use the entire shield and by definition the fields penetrate to the inside. Unfortunately, the weave still invites currents to penetrate the shield at high frequencies. This penetration occurs despite skin effect, as the strands of wire undulate between the two surfaces. Two shields are an improvement over one shield, as the parasitic capacitance is greatly reduced. Because the shields are tightly coupled to each other, shield currents still penetrate the weave at high frequencies. The only way to avoid penetration at high frequencies is to use a solid covering.

Another method of shielding is to provide a single wrap of aluminum foil around the signal conductors. This foil wrap is often placed around a two-wire twisted pair. This foil is folded over at a seam to stop field penetration. The aluminum foil is a good analog shield, but it has mechanical disadvantages. The aluminum foil tends to break if there is a lot of bending near a connector and soldering to aluminum is not practical. A drain wire that makes contact with the inner surface of the shield is added to the cable. The drain wire is used for a connection to the shield at both ends of the cable.

This solution has high-frequency problems. The drain wire carries ground currents on the inside of the cable. This couples fields directly to the signal conductors. If this cable is used for digital transport, the interference can be severe enough to cause logic errors.

One side of the aluminum foil is anodized. The anodized surface is usually placed on the outer surface to help avoid corrosion. If the anodization is removed and the drain wire is in contact with the shield on the outside surface, some of the high-frequency coupling problems might be lessened. The seam in the foil is not perfect, and field energy may still penetrate the shield. The foil is irregular and allows for distributed reflections. For short runs many of these imperfections can be ignored.

The cable television industry has the problem of transporting many channels of information over long distances. The eye is very sensitive to any form of reflection or delayed absorption. These processes appear as ghosts. To avoid what is called *distributed dielectric absorption*, the preferred dielectric is air. The center conductor is a stiff conductor centered in the cable by beads or a nylon cord. This cord spirals around the center conductor. The shield is usually extruded aluminum with a near-mirrorlike inner surface. This avoids reflections caused by minute variations in the characteristic impedance. In analog applications the drain wire can add high-frequency common-mode interference to the signals. Designs often have small passive filters added to the input leads to remove this interference (see Section 5.4).

The twisting of signal pairs in a shielded cable is helpful in canceling low-frequency normal-mode (differential) magnetic field coupling. Note that the sense of the coupling is reversed for every half-twist. Twisting is not used in cables where the characteristic impedance needs to be controlled. Because the twists are not tight (five or six twists per foot), magnetic coupling may occur when a cable runs near a magnetic field source such as a local power transformer. In this geometry the intense portion of the field is near one of the twists and there is no cancellation.

## 6.13  TRANSFER IMPEDANCE

Cable manufacturers often provide a measure of cable performance known as *transfer impedance*. A section of cable is terminated at both ends in its characteristic impedance. A current is caused to flow in the shield and a voltage is measured at the cable termination. The field energy that couples to the inside of the cable is transported equally in both directions. Since the cable is properly terminated there are no reflections. The ratio of twice the terminating voltage to the current on the shield is called *transfer impedance*. This circuit is shown in Figure 6.5. This impedance is adjusted for a cable that is 1 m long. The transfer impedance varies with frequency, so a curve must be supplied with each cable type. In the spirit of WCC, the coupling is assumed proportional to cable length up to one half-wavelength at the frequency of interest. Above this cable length the coupling is assumed to be no greater than the half-wavelength figure.

As an example, assume that the transfer impedance of a cable is 0.1 $\Omega/m$ at 10 MHz. This means that a shield current of 1 A in a 1-m-long cable will couple 0.05 V at the terminating characteristic impedance. If the cable is 2 m long, the voltage doubles. This assumes that the half-wavelength in free space is greater than 2 m.

The transfer impedance of most braided cable rises with frequency to a maximum around 100 MHz. At this frequency most braid is ineffective as a shield. To lower the transfer impedance, the shield must be made from some form of solid conducting tube. These cables, often called *thin-wall*, can limit

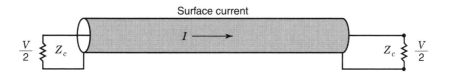

Transfer impedance $Z_T = \dfrac{V}{I}$ per meter

**FIGURE 6.5**  Transfer impedance.

the transfer impedance to milliohms at 100 MHz. Thin-wall shield is often corrugated to improve flexibility.

## 6.14   MECHANICAL CABLE TERMINATIONS

Shielded cables that interconnect pieces of hardware are often subject to a harsh electrical environment. The connectors and the termination of shields are often the main source of interference coupling. The shield current should ideally be kept on the outside surface of the shield. If the braid is bared back and twisted for termination, the shield current must concentrate in this connection. Any concentration of shield current implies an increase in both inductance and local field intensity. This is the point where the field enters the cable and propagates via the conductors and connector into the hardware.

The shield should terminate around the perimeter of the connector. In larger cables the connector might require a back shell. It is incorrect to bring the shield inside the hardware, as this invites interference to enter the hardware. The proper way to terminate a shield is to provide a smooth transition for current flow from the outside surface of the shield to the conductive housing. This is exactly opposite the treatment of a guard shield in analog applications.

## 6.15   PROBLEMS

1. The capacitance between conductors in a ribbon cable is 5 pF/ft. A 10-V, 10-MHz clock signal on one conductor has a rise time of 10 ns. An adjacent conductor has an impedance level of 100 Ω. What is the voltage coupled to this conductor if the cable length is 3 ft?

2. A 10-conductor ribbon cable has a ground return conductor on one edge and a 10-V clock signal carrying 20 mA on the other edge. The rise time is 10 ns. The conductor spacings are 0.05 in. The cable length is 50 cm. What is the radiation level at 10 m?

3. In problem 2, assume that the ground return is an adjacent conductor. What is the radiation level at 10 m?

4. A cable has a transfer impedance of 0.2 Ω/m at 50 MHz. The outer shield current is 100 mA at this frequency. What is the voltage coupled to the unterminated end of the cable if the cable length is 2 m. The cable is terminated at the other end.

5. A ribbon cable is routed between two circuit boards. The area between the ribbon cable and the conductive housing is 300 cm². The electric field in this area is 10 V/m at 20 MHz. What is the common-mode voltage that is coupled to this cable?

**6.** A circuit card has 500 traces where the average trace length is 4 cm. The spacing over the ground plane is 2 mm and 25% of the traces transition at each clock time. If the circuit uses 5-V logic and the rise time for the logic is 0.03 $\mu$s, what is the WCC radiation at 3 m? Assume an impedance level of 100 $\Omega$ per trace.

**7.** In problem 6, the loop areas for the power connection to the ICs is 0.3 cm$^2$. There are 50 ICs. Assume that on average, 10 gates transition per clock time per IC. What is the radiation from these loops at 3 m?

## 6.16 MOUNTING POWER TRANSISTORS

Power transistors are used in many switching applications. The current levels require that the transistors be mounted on large conductive surfaces. This is the only way to dissipate the heat generated in the transistors. The case of the transistor is often the collector, which precludes mounting the device in ohmic contact with the conductive surface. The transistor is often mounted using an insulating washer or gasket.

The transistor and an insulating gasket have a capacitance to the conductive surface. The current in this capacitance can be a source of interference, as the return path can have a large loop area. If an outside mounting surface is the equipment ground, the return path might involve conductors in the entire facility. Mounted on an inside surface, there is a chance to contain the fields.

One technique to limit current flow in an equipment ground is to use a guard gasket. This gasket has a center conductor that is insulated on both sides. The leakage capacitance around the gasket can be held to a few picofarads. The guard conductor returns most of the parasitic current to the circuit. This circuit arrangement is shown in Figure 6.6.

Power transistors used at high frequencies are often mounted directly on a power plane or ground plane. The manufacturer often supplies the geometry for this connection. This is necessary to limit inductance and common-impedance coupling so that the device can meet its specifications. If the case of the device is shown bonded to the ground plane, this advice should be followed. Rivets or even a gasket may prove to be an inadequate contact.

## 6.17 ELECTROSTATIC DISCHARGE

Electrostatic discharge (ESD) is the result of charge buildup. When the voltage gradient exceeds the breakdown potential of air, an arc results. A typical breakdown voltage figure is 1 kV/mm. Charge buildup is often the result of friction, whether it is between dust or grain particles in a grain elevator or a person walking on a rug. The charge buildup is most apt to occur when the humidity is below 30%. Above this humidity level, the charges can dissipate as rapidly as they can collect.

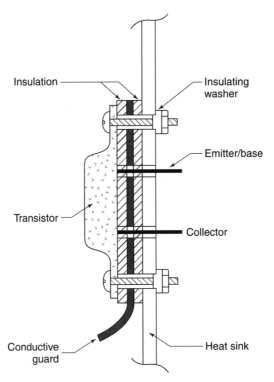

**FIGURE 6.6**   Guard gasket applied to a switching transistor.

The voltage-to-charge relationship is simply capacitance. The capacitance to ground for a human being is about 300 pF. A 1-s time constant for discharge involves a resistance of 3300 MΩ. To keep charge from building up on a human being, the floor can be made slightly conductive. This cannot remove charges that might accumulate on clothing as a result of local friction.

When a person with a charge approaches a grounded conductor (key to doorknob or finger to control switch), the charges on the body reconfigure and concentrate near the point of contact. The electric field and the resulting field intensity are greatest in the closing gap. For a human being about to touch a knob, the field energy is stored mainly in a volume about 2 ft in diameter. The rise time of the current pulse corresponds to the time it takes wave energy to travel 1 ft. This time is about 1 ns. The frequency of interest is $1/\pi\tau$, or about 300 MHz. The charge buildup on 300 $\rho$F for 10,000 V is 3 $\mu$C. The energy that is released goes into heat, ionization, light, and radiation. The field that radiates can be assumed to originate from a 5-A pulse. The interface distance for this frequency is 16 cm. The $H$ field at this distance is about 5 A/m. The $E$ field at this distance is over 1800 V/m. This is a significant field that can easily penetrate nearby apertures.

## 6.18   ESD PRECAUTIONS

Semiconductor components are vulnerable to ESD. They are usually shipped in conductive packages to limit any charge buildup. When the components are removed from these packages, care must again be taken not to introduce a further ESD problem. Assembly people are often grounded through a conductive strap to limit any charge buildup. Dress codes are enforced so that certain materials are excluded. The best preventive measure is humidity control.

If a charge should build up on a semiconductor product, bringing a ground near the product can cause an arc. The problem is complicated by the fact that minor damage may not show up in testing and the damage may limit the life of the product.

The most vulnerable points in a circuit are often at a connector (input, output, or control). This is a place where zener diodes can be used to clamp the leads to prevent an ESD pulse from doing damage. A small series resistor should be used to limit the zener current. If care is not taken in the remaining design, damage can occur anywhere there is a large enough loop area. Protective circuits must be placed at the points of entry, not inside the hardware on a circuit board.

If the ESD current path involves the equipment ground, it is wise to provide a wide conductive path away from the electronics. This could be a strip of aluminum foil pasted on a plastic panel. The field should follow the space away from the hardware, not on the inside near the electronics.

The ESD testing of floating hardware can be difficult. Repeated pulsing can lead to static charge buildup. At some point in time there will be a voltage breakdown that could damage the hardware. If there is a connector involved, a leakage path of 1000 M$\Omega$ is all that is required to bleed off this charge. This is the resistance of a gummed paper label.

## 6.19   ZAPPING

Testing devices called *zappers* are available that generate intentional ESD pulses. ESD testing before product release is important, as products do sustain ESD hits in the field. There are further uses for ESD testing. Many pieces of hardware must meet military specifications that limit radiation levels. There is a close correlation between susceptibility and radiation. If a product is insensitive to ESD tests, it probably will pass all radiation tests and not be susceptible to external radiation.

Zappers often have two modes. In the first mode, arcing occur at the probe tip. In the second mode, the arc takes place before the probe and a pulse of current is injected at the point of contact. The arcs can be varied in intensity and in frequency. The testing involves injecting pulses at all key points, such as at controls and cable entries. The testing starts out at a low voltage, as the intent is not to destroy hardware. If problems develop, some change is needed in design before the voltage is raised. The critical voltage range is at about 6000 V. The arc at the probe tip is a more severe test, as radiation from this arc is much closer to possible apertures.

## 6.20   PRODUCT TESTING: RADIATION

Electronic products that are sold commercially must meet radiation limits set by the Federal Communications Commission (FCC). These regulations relate primarily to radio and television interference. Certified testing requires a large room that is relatively free of external radiation. The *device under test* (DUT) is placed on a rotating grounded surface and a series of antennas at a fixed distance measure the radiated field strength. The antennas are calibrated and connected to a spectrum analyzer. The entire area is on a ground plane that is usually some sort of wire mat. All the hardware is bonded to the mat. The spectrum analyzer uses a moving filter that allows the field strength to be measured over a wide spectrum. The bandwidth of the filter is varied, depending on the part of the spectrum being measured. The bandwidth might be 10 kHz at 1 MHz and 100 kHz at 10 MHz. The table is rotated for a maximum signal. The unit passes the test when the worst-case radiation is below the prescribed limits.

Certified testing is expensive. Many laboratories have their own testing facilities that allow for a preliminary product evaluation. A testing facility will rarely recommend product modifications if a test fails. Obviously, repeated visits to a testing facility should be avoided.

The radiation from digital circuits does not always show peaks of radiation at exactly the clock frequency or of its harmonics. The reason relates to the numerous reflections that take place on all the traces. None of the traces are terminated in their characteristic impedance. There is spectrum that depends on path length. Another complication results because the trace-terminating impedances are nonlinear. The result is a smearing of frequency content. To further complicate the issue, the analyzer always looks at a narrow band of information, not at individual frequencies.

## 6.21   MILITARY TESTING

The military have there own testing standards. These specifications involve susceptibility as well as radiation limits. An enemy can try to knock out com- munications or navigation systems by transmitting large interfering signals. An aircraft must be able to fly directly over a powerful transmitter and not malfunction. Cabling between pieces of hardware may be subject to noise from other wiring. Testing for susceptibility is far more complex than testing for radiation. The problem includes power line susceptibility, radiation from

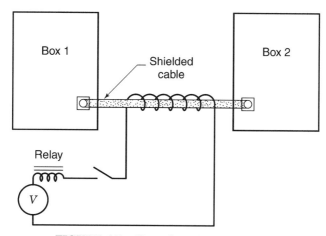

**FIGURE 6.7**    Chattering relay test circuit.

a variety of field sources, and system evaluation. Providing intense fields in a controlled region is expensive. Measuring performance while the equipment is under test poses its own problems. The zapping tests described in Section 6.19 can be quite valuable in preliminary evaluations.

## 6.22    CHATTERING RELAY TEST

The military has a test that places interconnecting cables under great electrical stress. A relay coil in series with its own contact forms a buzzer. The relay wiring is coiled around the cable. The number of turns is proportional to cable length. The relay contacts arc, and this introduces very high frequency noise. The coils of wire act as a step-down transformer, inducing current into the shield loop. The noise spectrum is very wide. If the shield is not terminated correctly, the field entry into the hardware at the connectors can be very disruptive. This test circuit is shown in Figure 6.7. See the comments in Section 6.13 regarding cable terminations. A test that is just as severe is to operate a handheld drill near the hardware. It is a real challenge to design hardware that is insensitive to this type of interference.

## 6.23    EURO STANDARDS

Products sold in Europe must meet a set of standards endorsed by the common market. Many of these requirements parallel an earlier set of German standards. Euro standards are more extensive and more difficult to meet than

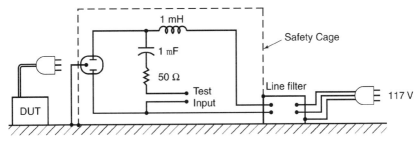

**FIGURE 6.8**  LISN circuit specified by the FCC.

the FCC rules in the United States. Manufacturers that intend to sell products worldwide will opt for the Euro endorsement.

## 6.24  LISN

The line impedance simulation network (LISN) is specified by the FCC for line testing. The network decouples the power line from the hardware at high frequencies. This decoupling allows pulses or sine waves to be superposed on the power line at a reasonable impedance level. The hardware can then be monitored for performance during this test. The inductor should be an air core device, so that it does not saturate. The diameter of the coil should be large enough to keep the resonant frequency above 10 MHz. The wire size should handle at least 5 A of line current. Commercial LISNs are available, but the unit can be built in any electronics facility (Figure 6.8). It should be totally enclosed, as the inductor is at the power line potential.

## 6.25  SNIFFERS

A *sniffer* is a very simple device that can be built to sense leakage fields. A loop is formed at the end of a shielded cable. The center conductor connects to the shield, forming the loop. This geometry is shown in Figure 6.9. The shield limits the *E* field so that the loop does not function well in free space. The device can be used to locate apertures that are radiating. If the loop is placed close to a conductor carrying current, the *H* field will be sensed.

The cable is connected directly to the input of an oscilloscope. The field strength can be calculated using the loop area and Faraday's law. An exact calibration of the sniffer is not required. The relative magnitude of the signal will identify the problems that need attention. The standard oscilloscope probe can be used as a sniffer. The prope tip is connected to the ground clip and the loop is placed near suspected leakage fields. The sensing loop is not shielded, but it can be effective in locating leaks. The probe makes no contact with the circuit.

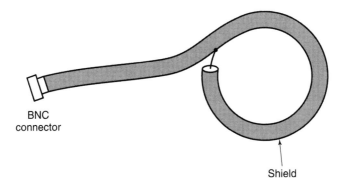

**FIGURE 6.9**   Sniffer.

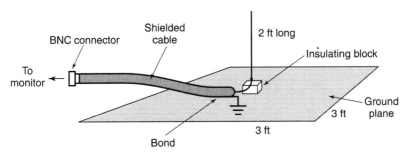

Support wire in a block of wood glued to ground plane

**FIGURE 6.10**   Simple antenna for field measurement.


## 6.26   SIMPLE ANTENNA

Testing for radiation involves a variety of antennas. The simplest antenna is a half-dipole (Figure 6.10). This can serve to monitor a piece of equipment for radiation up to about 20 MHz. The antenna is a vertical conductor supported over a ground plane that is about 1 m on a side. A rectangular sheet of metal will serve. The signal is sensed between the root of the antenna and the ground plane by a coaxial cable. The voltage is the $E$ field times the ratio of the antenna length to the half-wavelength of the radiation. Care must be taken that the field is from the hardware, not the environment. It is a simple matter to turn the hardware off and note the loss of signal. Again, this is not a well-calibrated procedure. On a relative basis it can be used to diagnose trouble.


## 6.27   PERIPHERALS

Manufacturers must certify that their equipment meets certain radiation stan-dards. This certification extends to monitors, printers, scanners, and modems, to mention a few hardware types. There is no guarantee that an interconnected set of hardware will meet the same specification. No one manufacturer is held responsible, as no one knows which peripherals will be used. The cable lengths and the external power connections are not under any type of control. Testing individual pieces of hardware is better than no testing, but often it can fall short. The loops created by interconnection can add to the radiation problems. If the loop areas are large, the equipment can be susceptible to external fields. The FCC specifically avoids susceptibility issues. Their only worry is radiation.

This issue on a larger scale is called *systems integration*. Examples include aircraft and automobile design. It is very difficult to test an entire aircraft in a radiation-free environment. Individual components can pass, but in a system environment, there can be problems. This is where the engineer can be severely challenged.

## 6.28  PROBLEMS

1. A sniffer has a loop area of 100 cm$^2$. The signal at the oscilloscope is a 0.2-V sine wave at 2 MHz. What is the $H$ field at the loop?

2. An ESD pulse occurs 10 cm away from an aperture that is 1 cm in diameter. What is the WCC for the $E$ field on the other side of this opening? Assume plane waves.

3. Hardware is hit by an ESD pulse. At a connector 10 cm away the shield opening is 2 cm long. What is the normal mode coupling into the cable if the conductor spacing through the connector is 1 cm$^2$? Treat the connector opening as an aperture.

4. In problem 3, what sort of $RC$ filter is required to limit the peak voltage to 0.2 V?

## 6.29  LIGHTNING

Lightning strikes are frequent events for many areas of the world. Lightning is the result of charge buildup in the atmosphere. When the voltage gradient exceeds the breakdown voltage, accumulated charge is transferred by arc to the earth. A great deal of lightning activity occurs cloud to cloud. The greatest damage occurs from a direct hit that enters a facility.

At most times there is a voltage gradient in the air at the surface of the earth. The equipotential surfaces near the earth conform to the shape of the conductors on the surface of the earth. The gradient is highest near vertical objects such as trees or flagpoles. During storm activity the frictional forces of wind and rain add to the ionization and the gradient can increase significantly. The gradient is highest near pointed objects.

Lightning starts when the air breaks down in a small region of space. A small amount of charge uses this ionized path to discharge the stored energy in the local capacitance. This breakdown results in a higher voltage gradient in nearby space. This results in the propagation of arcs, which further increases the gradient. This filament of current flow will extend from earth to a region of charge concentration in a cloud, where it branches out. The ionized paths act like conductors, and the cloud discharges a large pulse of current through to earth or a nearby cloud. This is the main lightning strike.

The volume in space that is discharged into the main ionized path can extend hundreds of feet around the starting arc. The region of discharge looks like a long cylinder. The arc path length in the cloud can vary, and this means that the current amplitude can vary.

The rise time that is usually attributed to a lightning pulse is about 0.50 $\mu$s. This means that the air cylinder supplying charge to the pulse extends about 500 ft around the initial streamers. The frequency assigned for a WCC is 640 kHz. The current in the pulse can vary to a maximum of about 100,000 A.

It is interesting to use these two numbers to calculate the voltage drop along a round conductor that is 1 m long. The inductance is about 1 $\mu$H. The reactance at 640 kHz is about 4 $\Omega$. A current of 100,000 A represents a voltage drop of 400,000 V. To reduce this voltage, many parallel paths are required. On a steel girder in a building the inductance per meter is not much different. On a direct hit the voltage drop per floor can easily exceed 1 million volts.

## 6.30  PROBLEMS

1. A lightning pulse of 50,000 A flows down a steel girder. A computer in the building is located 1 ft from the girder. What is the common-mode voltage between a ribbon cable and a rack panel where the loop area is 200 cm$^2$?

2. A lightning down conductor goes over a small wall that is 6 in. thick. The inductance of the path is 20 $\mu$H. The breakdown voltage in air is 50,000 V/in. Will a 20,000-A pulse arc through the concrete?

## 6.31  MIXING ANALOG AND DIGITAL DESIGN

Analog signals are often brought to a circuit board for conversion to a digital format. A question is frequently raised whether there should be a separate ground for the analog portion. The concern is that the digital processes will contribute interference to the analog signal. Analog-to-digital (A/D) converters often have more than 12-bit resolution, or 1 part in 4000. For a 5-V signal this is an error of about 1 mV.

The two ground planes require a connection. If this connection is a trace, currents that flow between the two areas must "neck down," and this is an inductance. The resulting voltage is just the interference that must be avoided.

The best way to avoid difficulty is to separate the fields for each function. Locate traces, connections in connectors, and components so that their fields do not share the same space. The currents will not share the same portions of the ground plane, and there will be no crosstalk from this common impedance.

Many A/D converters have their own differential input. If they exist, the ground potential differences are common-mode voltages and the converter rejects this mode. The A/D converter must be connected according to the manufacturer's specifications.

## 6.32  GROUND BOUNCE

The term *ground bounce* has appeared in the literature and implies that the ground plane has a voltage drop. The problem with this viewpoint is that all voltage is field phenomena. If large currents are concentrating in a region, there will be large fields. A voltmeter between points will sense this field by coupling to a loop area. The expression itself implies a misunderstanding of what is taking place. If the voltmeter (oscilloscope) is placed correctly, the bounce may disappear.

## 6.33  REVIEW

Most digital circuits are mounted on circuit cards that incorporate ground planes. The transport of logic signals is handled by traces that are actually short transmission lines. These lines require energy to support the transmission. This energy should be supplied locally be decoupling capacitors. If this energy must come a longer distance, the logic may malfunction and there may be radiation problems.

Many problems can occur when logic signals leave a circuit card on ribbon cables or on connectors. Nearby return paths should be provided to avoid radiation and cross coupling. The ribbon cable should be routed along ground planes to avoid common-mode coupling. Shielded cables can be used for the transport of analog or digital data. For coaxial performance over long runs, the cable quality must be considered carefully. The selection of cable type requires an understanding of the environment, the frequencies involved, and how the cables are terminated.

There are many regulations that must be met depending on product area. There are simple tools that can be built to aid the engineer in the early stages of design. Certifications of compliance are expensive. The military requirements are the most extensive. Products intended for overseas sales must pass standards that are tighter then those of the FCC in the United States.

ESD can produce a significant electromagnetic field. If this field enters hardware, damage can result. Lightning is the result of charge accumulation in clouds. The potential gradient between the cloud and earth gets high enough so that the air is ionized. The current levels in a lightning pulse can be as great as 100,000 A.

# 7 Facilities and Sites

## 7.1 INTRODUCTION

Electronic hardware can be found in almost every facility in the world. The character of the electronics varies over a wide range. The design and construction of aircraft requires complex testing bays. Communication systems use many computers, switching networks, transmitters, and receivers. Hospitals have monitoring and testing hardware as well as special safety considerations. A factory may have automated assembly lines, variable-speed conveyer belts, as well as temperature, mass flow, and pressure monitoring. A power generating plant has monitors, sensors, recorders, and signaling devices. There are ships in the navy, commercial aircraft, automobiles, police systems, and Federal Aviation Administration (FAA) facilities. The list goes on and on. Most of these facilities or sites use utility power with some sort of auxiliary power backup. All of these facilities operate many pieces of interconnected electronic hardware, as well as lights, air conditioning, and motors.

Research facilities and design groups have an opportunity to see their own products evolve. The majority of electronic hardware is purchased off the shelf or purchased from a company that designs and manufactures to specifications. The end users connect the hardware into a system and trust that the designs are compatible with their operating requirements. Often, corrections and modifications are needed to make the system function. The problems can relate to the environment, power line interference, cabling and interconnection difficulties, or simply incompatibility between devices. Sometimes the difficulty relates to utility power distribution or to some sort of interference. These problem areas are often the domain of the consultant.

## 7.2 UTILITY POWER

The power utility grounds its neutral or grounded conductor at each service entrance. (Grounding here means a connection to earth.) This practice stems from the need to provide an earth path for lightning that might strike the power line. Lightning is not apt to enter a facility on power conductors when they enter a facility in underground conduit. The neutral or grounding conductor for a distribution transformer that is located outside a facility is also earthed. If several facilities are connected to the same distribution transformer, there are further connections to the earth at each service entrance. Inside each facility the grounded conductor or neutral conductor cannot be earthed or grounded again. A second connection can affect the fault protection system.

The neutral connection to earth is for lightning and safety protection. The path should be direct and not circuitous, as this adds inductance to the path. Multiple earth connections outside a facility and the fact that there is neutral current flow implies that there will be power current flowing in the earth. As mentioned before, this current follows building steel and any conduit that contacts the earth. This current and its associated fields enter every facility.

In a facility without power distribution transformers, one neutral or one grounded conductor is supplied to most loads. There are voltage drops in the neutral or grounded conductor, implying that it is not at the same potential as the earth connection at the service entrance. With typical electronic loads the nonlinear currents that flow in the neutral have harmonic content that adds to the neutral voltage drop. It is not uncommon for the voltage drop on the neutral or grounded conductor to be several volts. The grounded conductor or neutral conductor may not be lifted from earth for any reason. It is unsafe and unlawful to do so. Ungrounded power systems are allowed but only under special circumstances.

The rules for power distribution are controlled by the National Electrical Code. Local municipalities pattern their regulations after the NEC. It is vital for the safely of others that these codes be followed. These codes are the result of years of experience. The reasons behind some of the practices are obscure. No rationale is given in the code. It is important that everyone follow the same rules. Ten years from now a person making changes to a nonstandard system might unintentionally create a hazardous condition that could harm personnel or start a fire.

Residential power and many small facilities are supplied power from 120-V single-phase conductors. The utility transformer often has a grounded center-tapped leg that is the grounded conductor. The four conductors are 120 V, 0 V, 120 V, and the equipment (green wire) ground. The voltage is 240 V between the two hot conductors. Larger facilities are supplied with three-phase power. This power is stepped down through various transformers for lighting, office loads, and motor loads. Often, a separate transformer supplies power to electronic loads to avoid transient coupling from air conditioning and various motor loads. The three-phase voltages on a wye-connected secondary might be neutral, the three three-phase lines at 208 V and a safety wire (equipment ground conductor).

On-site distribution transformers are called *separately derived sources of power*. The secondary neutral is grounded just like a service entrance. This neutral is connected to the nearest point on the equipment grounding system. In larger facilities this is often building steel. There is a breaker system for all loads, and this is the point where all equipment grounds associated with the secondary are returned and earthed. This method of power distribution provides a separate neutral for a set of loads and avoids the voltage drop in

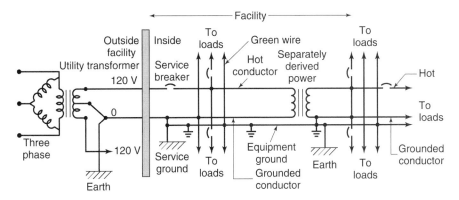

Note: Only one phase is shown. ⌐◠⌐ is the symbol for a circuit breaker.

**FIGURE 7.1** Power distribution system.

the neutral used in the rest of the facility. A typical separately derived power source is shown in Figure 7.1.

Grounding one of the three-phase legs in a three-phase transformer is not allowed, as this unnecessarily increases the ground-to-line voltages. The code requires that grounding should always be handled to keep line voltages to ground to a minimum. Grounding of a delta leg at its center tap is permitted. The only constraint is that the high-voltage leg must be specially color coded. In this case there is no neutral point to cause confusion.

## 7.3  FLOATING UTILITY POWER

If the secondary of a power transformer were left floating (unearthed), the voltages with respect to earth would wander until leakage resistances and leakage capacitances would reach some balance. Some of these resistances might be nonlinear, resulting in rectification. One way to limit this wandering would be to place two zener diodes back to back in series from one of the lines to earth. If the voltage exceeds the zener rating, the average voltage would be clamped.

Floating systems are permitted when there is a qualified electrician available. An application might be an electric crucible. If there is a fault from the wiring to the crucible housing, the power cannot be turned off or the crucible will be lost. The proper procedure is to sound an alarm, finish using the crucible, and repair the crucible off-line.

Floating systems are used on board naval ships. This avoids problems of corrosion. If there is a simple fault during battle, the ship is still supplied with power. A portable generator can be operated without an earth connection as long as the power is grounded to the generator housing. A floating system is a very noisy power source for electronics. The reason relates to the switching and operation of other loads. Switches rarely close on zero voltage crossings. When a switch closes, the effect is like a rectifier. The power conductors all change voltage in a step manner. This reactively couples noise directly across each hardware transformer. Nonlinear loads also affect the neutral potential. This noise also couples to the secondary across the hardware transformers. The proper solution is to add a separately derived transformer to power sensitive electronics. This added transformer has a grounded neutral or a grounded secondary. Now the floating system cannot couple neutral-related noise into the hardware. The ship still has floating power.

## 7.4   ISOLATED GROUNDS

An *isolated ground* is not a floating ground; rather, the term refers to how and where the equipment grounding conductor (green wire) is grounded. In normal wiring all receptacles are connected to the equipment grounding conductor. Also, all receptacles are bonded to each other via any interconnecting cable or conduit. In an isolated grounding arrangement the equipment grounding conductor for each receptacle is wired separately back to a subpanel or possibly back to the service entrance. It is isolated from any other connections. The receptacles are still grounded normally. An isolated receptacle must be specially identified for this application.

The original intent in supplying isolated grounds was to avoid sharing a common impedance in the grounding of electronic hardware. Line filters for hardware normally return power line noise current via the equipment ground. If a second piece of hardware shares a conductor, the filtered noise would cross-couple into this hardware. This was felt to be a source of difficulty requiring correction. Bringing all the equipment grounds to one point agreed with the philosophy of single-point grounding. Unfortunately, this idea cannot work at high frequencies. The inductance of a long equipment ground connection negates the performance of the line filter. Line filter currents then find a new path to take. If there are hardware interconnections, the noise currents will use these paths. The source of this cross-coupling interference can be difficult to locate. The best systems approach is to use the equipment grounding conductors as a quasi ground plane. There are many conductors, all connected together in a gridlike structure. This is a low-impedance grid for noise currents and the filters can function. Ideally, if all the hardware were bonded to a valid ground plane, there would be no common-mode field created by filter currents that would couple into interconnecting cables. This is an example of how many connections (ground loops) can solve a problem.

This problem of isolated grounds is not severe when the equipment is all bonded to a common rack. The greatest difficulty arises when the hardware is stand-alone or housed in separate racks. An example might be two computers powered from separate receptacles. The loop that is formed by the isolated equipment grounding conductors can couple to external fields. The results are common-mode voltages impressed on interconnecting cables. These voltages can burn out components. It is important to verify that the receptacles involved are bonded together by conduit, as this limits the loop area.

## 7.5  SINGLE-POINT GROUNDING

In the early days of electronics, engineers found out how to build sophisticated pieces of hardware. The hardware dimensions were limited to a maximum of 36 in. and the bandwidth did not exceed 50 kHz. The trick was to use single-point grounding. A stud was mounted to a metal housing and every ground in the circuit was brought back to this point. The sequence of connection was determined by signal sensitivity. High-level signal grounds were kept separate from low-level signal grounds. This approach was successful and it began to appear in larger systems with more bandwidth. Radio and television hardware designers never joined in this approach, as they had already developed high-frequency techniques. Single-point grounding methods would not have worked had they been tried.

The days of aerospace development and of building defense facilities raised issues of how to ground things. The single-point idea prevailed. Shields from many analog signals were grouped together and single-point grounded. This approach had its failings but few alternatives were suggested. This idea failed miserably in digital circuit development and the world turned to ground planes for circuit boards. There was a recognition that ground planes were needed, but how they worked was not explained. As a result, many facilities were built with excellent ground planes but the facilities still had their problems.

One systems approach was to collect all the equipment grounds and carry them separately to a grounding well. This was the single-point grounding idea applied to facilities. Somehow this was accepted and written into various codes. To illustrate the resulting problem, consider the impact on a group of buildings. In one facility the utility grounding was local to each building. The secondary circuit grounds were all brought back to a single-point ground at a central well. There were ground planes in each building. When lightning hit in the area the potentials between the well and the service entrance grounds would exceed 30,000 V. This voltage appeared across facility transformers and some of them would blow up. Something was seriously wrong. Extending single-point grounding to buildings was a bad idea. Correcting the wiring would have been simple compared to the effort required to change the published codes.

Neutral grounding in facilities is a necessity and it is single-point grounding. Analog circuits can be built using single-point ideas, but other techniques are preferred. Common-mode rejection requires that each signal be guarded by its own zero of potential. Here the single-point connection for many shields fails. Digital circuitry and all radiating hardware cannot use single-point ideas. It

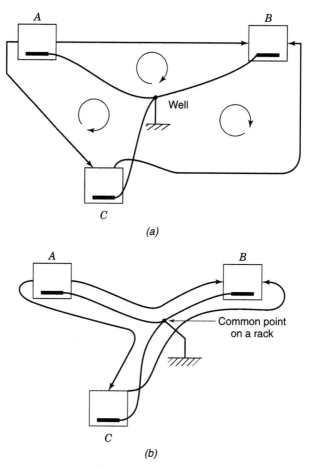

**FIGURE 7.2**  Star connection problem.

is clear that single-point grounding is not a fix-all technique. It is important to understand underlying principles and test to see how far any one approach can be taken. Following rules for rules' sake is not a good idea.

Single-point connections are often called *star connections*. The idea is to avoid any common impedance in the grounding system. Every circuit is grounded, but the loops do not share a common impedance path. This arrangement is shown in Figure 7.2*a*. The obvious difficulty is that loops are formed between every signal tie and the ground connections. These loops couple common-mode signals into every cable connection. In Figure 7.2*b* the loop problem is resolved by routing all cable interconnections along the grounding path. Usually, this is impractical. In a single rack of hardware this solution might be feasible. A single-point ground is selected and all cables are routed to and from this one point.

To some, grounding is a way to get rid of noise. The ground is viewed as a sump that collects noise current. This view runs contrary to any known physical principle. A ground or earth is just a conductor that can reflect or absorb field energy. The power system and nature use it for a lightning return. The earth is a boundary in most of the systems we operate. It must be used in an understandable and safe way.

It is illegal to have two distinct earthing systems in one facility. Consider a separately derived power transformer that is treated as a new service entrance. This new power source is earthed but makes no contact with the equipment grounds in the first system. If there is a fault from power in the new system to an equipment ground in the primary system, the earth becomes a load. This means that power potentials exist between equipment grounds in two parts of the facility. This is a severe safety hazard. A breaker may not trip, as the earth current may only be 10 A. The rule is simple. There can be one and only one equipment grounding system in a facility. All conductors that might contact power wiring must be bonded together and returned electrically to the service entrance.

## 7.6  GROUND PLANES

The ground planes used in digital hardware provide insight into their function. Traces and ground planes form transmission lines that confine fields. This approach is not practical in facilities. A ground plane can reflect field energy, but it cannot in any way absorb field energy. There are important uses for a ground plane in a facility that are different from the circuit board approach.

1. The ground plane provides lightning protection. If a pulse of current crosses the entire ground plane, the potential difference will be small. If the impedance of the plane is 500 $\mu\Omega$ per □ and the pulse is 50,000 A, the potential drop across the entire surface is 25 V. This is not going to cause damage.

2. The ground plane can provide ESD protection. If the floor tiles are slightly conductive, any charge build up on a human being will dissipate.

3. The ground plane can limit common-mode coupling. An $E$ field cannot exist parallel to the surface on a ground plane. If cables are routed next to the plane, common-mode coupling will not occur.

4. The ground plane is a reference surface for the racks that bond to its surface. This means that the hardware filters can use this common surface as their reference conductor. Currents from power line filters that flow on this surface will not generate a significant common-mode field.

5. The ground plane can be used as the bonding point on the equipment grounding conductor system for neutral grounding of a separately derived power source. This is the best way to avoid neutral voltage drops

that might exist in the rest of the facility. The new neutral will only handle currents from the local loads.

A ground plane provides lightning protection if the current enters the ground plane at many points. The potential drop near a single-point connection could be thousands of volts. Attempting to limit current flow by inductors is counterproductive. The National Electrical Code prohibits placing any component in the equipment grounding path. The installation should provide many connections from the stringers to building steel and to all conduit and rebars if available.

A ground plane often consists of bonded stringers that are on 3-ft centers. The stringer grid is supported on stanchions. Tile flooring is placed between the stringers. These tile surfaces are slightly conductive, to limit any charge buildup. Hardware that connects the stringers to the tile must be in place for the tiles to be effective. The low-impedance connections between stringers make the surface look like a mesh with large openings. This means that the aperture opening is that of one tile. This is not very effective, as the ground plane is not an enclosure. Note that the apertures around the perimeter of the ground plane are not controlled.

The common-mode coupling area is the space between cables and the ground plane. In most installations the cables are routed on the concrete floor under the ground plane. This means that common-mode coupling can still take place. Hanging cables on the stringers is better, but this is very inconvenient.


## 7.7  ALTERNATIVE GROUND PLANES

The tiles used with a stringer system are excellent for limiting ESD. The floor provides a way to hide and protect all the interconnecting cables. The surface can be a part of a plenum chamber for circulating cooling air to the hardware. Typically, this type of floor does not provide common-mode rejection for cables routed on the concrete floor. To solve this problem, a second ground plane should be present that rests under the cables and that bonds to the racks where the cables exit the racks. This plane can be a bonded wire grid or bonded sheets of metal.

The rebars in the concrete should not be considered a ground plane. There is no way to determine the existence of a bond at crossings. There is no way to know if the rebars bond to the building steel on the perimeter. Most important, there is no simple way to bond racks to the rebars. A ground plane without connections cannot be very effective.

Ground planes can extend between rooms through the use of many small conductors. For example, No. 10 wires on 6-in. centers can be fed through small holes in the wall and used to bond two ground planes together. Ground planes that should be continuous between floors should extend up the sides of a wall. At some point, any interconnecting cables must cross through the ground plane. It would be a real problem to extend ground planes between buildings. It is preferable to use fiber optics to interconnect signals where separate buildings are involved. Separately derived power for each building is also preferred.

## 7.8  POWER CENTERS

A power center is a separately derived transformer housed in a rack cabinet that can be bonded to a ground plane for powering other racks of electronic hardware. The power center contains circuit breakers, line filters, and possible surge protection. This arrangement provides a clean electrical environment for electronics mounted on the ground plane. Power centers are "listed" for this application.

The power center transformer has internal shields that limit common-mode effects. The shielding and filtering arrangement is shown in Figure 7.3. These shields are connected internally and not available for user definition. The center shield is bonded to the ground plane and limits power line common-mode signals from flowing to the secondary circuit. The primary and secondary shields are connected to the grounded conductors on their respective sides. There are mutual capacitances across the center shield. The primary and secondary shields return currents that flow in these capacitances to ground. Without these shields, this interference current would flow in the coils of the primary or secondary windings. These currents would then be converted to normal-mode signals by transformer action. As normal-mode signals they would appear on both sides of the transformer. This action is known as *common-mode to normal-mode conversion.*

The line filters in the power center take over when the shields are no longer effective. It is impractical to build transformer shields that are effective above

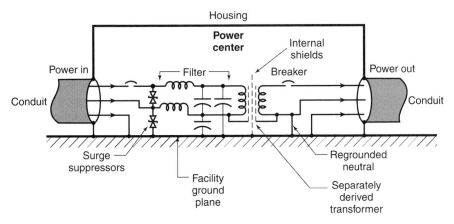

**FIGURE 7.3**  Power center.

about 20 kHz. Line filtering is feasible only if primary and secondary wiring are routed in separate conduits into separate enclosures. This is why secondary circuit breakers are located in the power center. Bringing power back to a common breaker panel would compromise the line filters. This construction keeps the shield lead lengths to a minimum, and this limits any fields that might be generated by shield current flow. Once shield current disperses on the ground plane, the associated nearby fields are essentially zero.

Because of proximity problems, the transformer in a power center must be designed for low-leakage inductance. To limit internal heating, the transformer must have a $k$ rating for handling high harmonic content in the load. Figure 7.3 shows the filtering and shielding for a single-phase transformer. For a three-phase transformer, there are shields and filters for each leg.


## 7.9  LIGHTNING PROTECTION

The charges that develop in a cloud create a voltage gradient to earth. When the gradient reaches a limit, a small region in the air ionizes. This is usually at or near a pointed object such as a tree or antenna. This ionization increases the nearby gradient and this adjacent region also ionizes. The result is an ionized path that propagates between the cloud and the earth. This connection or leader is used as a path to discharge the cloud. The current levels can be as large as 100,000 A. The rise time for the pulse is about 0.5 $\mu$s.

The method used to protect a facility is to provide a path to earth that the lighting is most likely to take. Sharp conducting objects are placed on the roof around the perimeter of a facility to provide a starting point for a leader. Down conductors are supplied to carry the pulse of current to earth. During weather the air is in constant motion and the gradient is not well defined. It is possible for the initial leader to start from a flat point on a roof. In other words, there is no guarantee that lightning will agree to strike specific points. The pointed objects placed on the building are called *air terminals*. An antenna that is higher than the air terminals can offer a better target. The antenna should be designed to accept a direct hit or the antenna should be removed.

At the moment of the strike the voltage at the tip of an air terminal can be millions of volts above earth. In other words, there is a significant electric field from this tip to the earth below. As the pulse develops, the gradient changes and this determines where the current will flow. The lightning circuit is a complex transmission line with electric and magnetic fields in transition around the lightning path. If the changing gradient can break down a shorter path through air, the arc will jump to another conductor. This can occur if the down conductor takes a sudden turn. A good example of this action occurs when lightning hits a tall building under construction. It is possible to observe diagonal arcing from horizontal steel beams to the vertical girders. Lightning will not follow the right-angle bend: Very simply, the path is too inductive. After the walls are in place, this arcing may not be visible. Major arcing inside a building is not an acceptable situation. The lightning paths should follow a vertical path without taking these diagonal paths.

Down conductors in a steel building are superfluous. The steel itself provides an excellent path. The current will not fan out at the lower stories. It will flow straight to earth in one of the steel members. If there is no steel in a building, a strike might follow air-conditioning ducts or conduit to find an earth path.

Down conductors that parallel a corrugated steel wall are also superfluous. The walls are the lowest-impedance paths; lightning will arc over to the wall rather than follow a down conductor. This large conducting surface spreads the current out, reducing the voltage gradient in this area. Air terminals on a steel roof are also superfluous. The current will jump to the roof and spread out, looking for parallel down paths. It is incorrect to place a hole in the roof and invite the pulse to find earth inside the building. If the roof is made of steel, the pulse will always arc to the roof. Inviting a pulse to go inside a building is a bad idea anyway.

Down conductors should be spaced away from a building to avoid arcing to internal down conductors. The recommended spacing is 6 ft. This may not look pretty, but it is a safe procedure. Down conductors that conform to the pitch of a roofline or that curve around an eave are not acceptable. Lightning will find some other path to take.

Ideally, there should be one down conductor for every air terminal. Connecting air terminals together to limit the number of down conductors adds inductance to the possible lightning path. This raises the possibility of arcing to nearby conductors such as air ducts or plumbing. Air terminals should still all be interconnected. These interconnecting conductors should transition around bends with a gentle radius to limit inductance. The recommended radius is given in the codes.

When the building has steel walls, the path to earth at the foot of the wall is important. A series of down conductors should be bonded to the wall and carried into the earth. If this is not done, the pulse may arc into the building at the foot of the wall and find another path to earth. In desert areas these points of earth entry should be kept wet so that the earth is conductive in this area. Arcing that occurs in the earth is not of concern.

The potential gradient on the earth's surface as the result of a lightning strike can be as high as 15,000 V/m. This gradient is sufficient to electrocute standing cattle in an open field. It is easy to see that cables running between buildings can couple to very high common-mode potentials if there is a nearby lightning strike. It is helpful to keep conductors inside conduit. The lightning field will cause current to flow on the outside of the conduit and not in the conductors. At points of cable termination, some sort of surge supression can be used to limit the voltage rise.

Facilities with many nearby structures should be connected by a grid, not by a few grounding conductors. This grid serves as a ground plane that limits potential differences between structures. Guy wires on large towers must be properly terminated. If lightning follows the guy wire to earth and arcing occurs at the foot of a guy wire, the tower might collapse if the wire melts. Arcing is necessary, but it should not compromise a point of mechanical stress. If the guy wire contacts steel in a concrete piling, the guy wire will be protected.

Telephone companies provide a special grounding gap (carbon blocks) for wiring where it enters a facility. If there is a lightning strike on the telephone line, excess voltage will cause an arc across this gap, allowing the pulse to enter the earth outside the facility. The hope is to reduce any line-to-earth potentials associated with the lightning pulse. Unfortunately, the earthing of utility wiring is often at a point separated from the telephone grounding gap. There can be a significant earth gradient during a strike, and the telephone line ground can be thousands of volts removed from utility ground. This can be a safety concern if a human being is associated with both conductors. Hardware that interfaces both power ground and the telephone lines must be provided with adequate surge protection to avoid blowing up circuits. A similar problem exists with cable television. To limit potential differences, the cable should enter a facility near the service entrance. The cable shield should be earthed at this entry point. This avoids another possible shock hazard during lightning activity. In some cases it may be possible to relocate cable entrances to reduce any risk of shock.

Connecting earth points together by copper wires is usually not effective in reducing potential differences cause by lightning. This is because the inductance of the wire connection is too high. A conductive ground plane is a possible solution, but this is not available in most situations. This is the reason for recommending that all electrical entries into a facility be localized. This definitely reduces the risk of excessive potential differences between various service conductors during lightning activity.

## 7.10   SURGE SUPPRESSION

A lightning pulse produces a significant field. This field can couple common- or normal-mode signals into hardware. Input circuits that might be vulnerable often have shunting components that absorb the energy and limit the voltage to protect the circuit. Typical components are zener diodes or metal oxide varisters (MOVs), which conduct over a certain voltage. In some circuits, diode clamps to power supply voltages can provide protection. The speed of response for lightning is significantly slower than for an ESD event. The speed of response is important in selecting the proper surge suppressor.

Surge suppressors can be placed between power lines and an equipment ground at the service entrance. This can be the first line of defense against lightning surges in a facility. Additional surge suppressors can be placed in breaker panels. Suppressor products are available commercially for this application. The large currents that can flow create an additional field that should not be near hardware. MOVs must be replaced after a limited number of pulses.

Commercial products often contain *LC* filters which slow the pulse rise time so that the MOVs can function properly. Surge suppressors for computers often provide an in–out connection for phone lines. Internal suppressors can limit the potential difference between the telephone lines and the equipment ground, thus protecting modems in the computer. Line suppressors should limit both normal and common-mode overvoltages. In hardware that is not rack mounted, common-mode pulses cannot be shunted properly.

## 7.11 RACKS

Racks for housing electronic hardware come with a variety of features. Features include cooling fans, power strips, and locking doors both front and rear. The cabinet may be gasketed to limit radiation from entering or leaving fields. Special hardware may be available to bond the racks to a ground plane.

In some applications the racks become the ground plane. Bonding racks together can be a problem. A series of screws between vertical members may not be adequate. Bonding requires clean plated surfaces that will not corrode. The bolts should use washers under spring pressure (Bellville washers) so that the contacts do not vibrate free. Surface currents should not have to take a circuitous route to get to all sides of the rack. The low impedances that are characteristic of a ground plane should be the rule.

Rack surfaces can be used to limit common-mode field coupling. This requires that all panels be bonded together to form a reference plane. Depending on how the rack is constructed, this plane may not exist. Some designers add a copper sheet on the rack floor. This sheet can cross between several racks. To be effective, this copper sheet must be bonded to the racks at each vertical wall. An unbonded sheet of copper is not a ground plane. Point connections are not satisfactory in bonding to form a ground plane.

Racks that are mounted on a ground plane should be located away from walls and vertical steel girders. This service space keeps the hardware away from any down conductors that might carry lightning. A further precaution is to avoid connecting metal ducts to the roof of the racks. If ducts are used, the last 6 ft should be plastic. This limits the possibility that a lightning strike can use the ducts and hardware as a path to earth.

## 7.12 MAGNETIC FIELDS AROUND DISTRIBUTION TRANSFORMERS

The leakage field around a distribution transformer can interfere with nearby computer monitors. The field modulates the electron beam and causes pattern distortion. People working in the area are often disturbed when they realize that they are spending many hours per day in this magnetic field.

This field is a low-impedance induction field that makes shielding very difficult. The leakage flux, usually the biggest offender, emanates around the transformer coils. One solution is to place a shield of high-permeability material around the monitors. This shielding reconfigures the field so that it avoids crossing the electron beam. The best material is Mumetal, which must be annealed in a magnetic field in a nitrogen atmosphere after it is fabricated. This is an expensive solution. The magnetic field is still present for the operator.

Placing a shunting magnetic path around transformer coils is also difficult and expensive. Before this is done, the field should be mapped. It is possible that a reorientation of the transformer might reduce the field in a critical area. Moving the monitors is another option. The shunting path solution is somewhat marginal, as the flux is difficult to capture. There is always an air gap in any added external geometry, which means that the effective permeability of the magnetic path is low. Steel in insulated sheets can be placed into position around the coils. In a big transformer this can be a very heavy structure that requires mechanical support. The added structure should also not affect the air cooling of the transformer.

## 7.13 MONITOR FIELDS

Magnetic deflection coils are used in monitors to move the electron beam. The fields from these coils extend around the monitor. These sawtooth fields move the beam horizontally and vertically. The deflecting coils are located around the base of the cathode ray tube (CRT). Improved designs limiting the extent of the field have been instituted as the result of health concerns. Providing shielding for the monitor is impractical because nothing can be placed in front of the tube except the viewer. Monitors that rely on thin-film technology do not have this deflection system.

The beam that strikes the front of a CRT monitor is dispersed at the surface. This dispersion radiates a small amount of energy. A pulsed beam also radiates a spectrum of information. A nearby antenna can be used to sense both radiation and deflection patterns. This information can then be processed to read the screen. This allows the information in a computer to be monitored at a distance.

## 7.14 MOTOR CONTROLLERS

Motor controllers function by connecting the power line for a fraction of the cycle. A gate on a triac controls the firing angle. The triac continues to conduct until the current falls to zero. Turning the triac off at a zero-voltage crossing requires that the current be forced to zero. The demand for energy at midcycle places a step demand for current on the line. These steps increase the harmonic content in the load current.

A rapidly changing magnetic field can exist in and around the motor. This field causes current to flow in equipment grounds. The field changes most rapidly at turn-on and turn-off times. Motor housings are not shields against magnetic fields. The field external to the motor will cause currents to flow in the external steel of the building. For large motors this interference can be significant. If the motor is mounted on an insulating pad, it is still grounded per code by the conduit. This will limit the opportunity for current to flow into the facility. If necessary, the shaft can be coupled via an insulator.

## 7.15   SCREEN ROOMS

A *screen room* is a large metal enclosure that limits the penetration of electromagnetic fields. The lack of external fields inside the room allows hardware to be tested for radiation. The other application involves secrecy. Hardware in the room can be operated without radiation leaving the controlled space. The word *screen* was used when screens or mesh were actually used. Today's screen rooms are made using thick conducting walls so that near-induction fields are properly attenuated.

The attenuation of fields by a screen room can exceed factors of 100,000. At frequencies where skin effect dominates, the thickness of the walls is more than adequate. Screen room design problems relate to power entry, human entry, and the operation of testing hardware. Other concerns include telephones, lights, and air conditioning.

The power entering a screen room provides an avenue for transmission going both ways. The line filters that are used are rather bulky and are mounted on the outside wall. The conduit supplying power should not strap to the room walls. This places fields associated with power (low-impedance induction fields) against the wall. The equipment ground connection is adequate for safety, and no other grounding conductors should be added. This addition could add surface currents that could penetrate the wall. The floor of the screen room is often near building steel. This steel may carry parasitic power currents. To limit coupling from this source, a double floor is sometimes provided.

There are many ways to provide human access. Finger stock or gaskets are used to close the door aperture. Waveguide principles are used to provide the needed attenuation. Air is supplied through a honeycomb filter. Any ductwork is insulated from the room to limit surface currents. Lightning poses no problem to the screen room interior. The standard 6-ft rule is still used to limit arcing to the room from any ducts.

Fluorescent lights radiate a lot of energy and should not be used in a screen room. Incandescent lights are required. The telephone lines need their own filters mounted to the screen room wall. If fiber optics are used, any steel support cable must be removed 6 ft from the wall. The best aperture to use for the optics is a long tube. This tube can be mounted parallel to the wall. Testing hardware can be on the outside of the room provided that cable shields are bonded to the wall and do not allow fields to cross the boundary in the wrong direction. Electrical access to the screen room should be limited to one area. This eliminates the flow of surface currents between entry points and improves the performance of the room.

Hardware should not be placed against the outside screen room walls. Inside the room it is important to avoid placing near induction fields next the walls. The corners inside the screen room are gusseted. Currents on the inside surfaces enter the wall rather than taking a sharp right-angle bend. The gussets are welded to the corners to provide a smooth transition for any internal surface currents.

## 7.16 REVIEW

Facility power in most cases is a grounded (earthed) system. This is necessary for lightning protection and human safety. The neutral is grounded once at the service entrance, an example of single-point grounding. In larger facilities, distribution transformers are used to supply voltages (feeders) for motor, lights, offices, and electronic loads.

Lightning damage can result if earth gradients are not considered. Cables provided to a residence should enter at the same point. Structures should be placed on ground planes rather than being connected together by single conductors.

Ground planes can be used to support a "clean" power environment for electronic hardware. The use of power centers that use a separately derived transformer can provide filtering and limit common-mode coupling. The ground plane can provide lightning and ESD protection as well as common-mode rejection. The transformers are shielded internally to limit power line interference.

Screen rooms serve to provide a clean environment for radiation testing. The rooms can also be used to maintain secrecy. There are many techniques that should be used in screen room installation and use so that the performance of the screen room is not compromised.

# APPENDIX I
## Solutions to Problems

### Section 1.15

1. 10 V/m; 10 V/0.1 m or 100 V/m; 10 V/0.01 m or 100 V/m; 10/0.001 m or 1000 V/m.

2. The voltage per centimeter is 0.2 V. The current in 10 kΩ is 0.2/10,000 A, or 20 $\mu$A.

3. The total series resistance is 2 kΩ + 3 kΩ, or 5 kΩ. For the parallel resistance, place 1 V across the two resistors. The current in the resistors is $\frac{1}{2}$ mA and $\frac{1}{3}$ mA; the total current is $\frac{5}{6}$ mA. The resistance equals 1 V divided by the current, or $\frac{6}{5}$ kΩ. (The reciprocal of *milli* is *kilo*.) The conductance is the current-to-voltage ratio: $\frac{5}{6}$ mΩ/1 V, or $\frac{5}{6}$ mS.

4. The voltage in 1 cm is 0.0001 V/m × 0.01 m = 1 $\mu$V. The resistance in 1 cm is 2 mΩ. The current is 1 $\mu$V/0.002 Ω, or 2000 $\mu$A, or 2 mA.

5. The potential difference is 5 V. This is the work per unit charge.

6. 1 mS equals 1000 Ω. 3 mS equals 333.3 Ω. The series resistance is 1333.3 Ω. For parallel resistors, their conductances add. The total conductance is 4 mS, or a resistance of 1/(0.004S) equals 250 Ω.

7. The 1000-Ω resistors in a square form two parallel paths of 2000 Ω each. This is equal to 1000 Ω. If 10 V is placed across one diagonal, the voltages on the other two corners are 5 V. The voltage difference is zero.

8. The voltage at the junction of the unchanged pair is 5 V. In the other pair the total resistance is 2010 Ω. The current is 10/2010 A. The voltage across the 1000-Ω resistor is 10/2010 × 1000, or 10/2.01 V. This is 4.98 V. The voltage difference is 5 V minus 4.98 V, or 0.02 V. The voltage across the 1010-Ω resistor is (10/2010) × 1010. This is 5.02 V. For this arrangement the voltage across the diagonal is 5 V − 5.02 V, or −0.02 V.

9. The internal resistance is the voltage drop divided by the current, or 0.1V/10 A, or 10 mΩ.

10. The charger must supply 12 V plus the drop across the internal resistance, or 12 V + 2 A × 0.01 Ω = 12.02 V.

**Section 1.18**

1. In the parallel circuit the dissipation in the 100-$\Omega$ resistor is $10^2/100$, or 1 W. The 200-$\Omega$ resistor dissipates $10^2/200$, or $\frac{1}{2}$ W. The total power is 1.5 W.

2. When the resistors are in series, the total current is $10/300 = 0.033$ A. The power is $0.033^2 \times$ resistance. For the 100-$\Omega$ resistor, this is 0.11 W. For the 200-$\Omega$ resistor, this is 0.22 W. The total power dissipated is thus 0.33 W.

3. A charge of 4 C in 8 s is a current of $\frac{1}{2}$ A. The power is volts times current, or $4 \times \frac{1}{2} = 2$ W. The energy is watts $\times$ time, or $2 \times 8 = 16$ J.

4. The resistors can all be placed in series or parallel. The third way is to form two pairs of parallel resistors and place them in series. The series group will accept the greatest voltage. The parallel group will dissipate the most power for the lowest voltage.

5. For $\frac{1}{8}$ W the voltage across each resistor is the square root of $100/8$, or 3.54 V. In series, the maximum voltage is 14.14 V. In parallel, the voltage is 3.54 V. For series–parallel, the voltage is 7.07 V.

6. 10 V for 2 s across 40 $\Omega$ is 2.5 W. The energy in 5 s is 5 J. 5 J in 5 s is 1 W. A 2-W resistor would be adequate.

7. The reversal of voltage does not change the wattage. The power is $24^2/48 = 12$ W.

8. The number of joules equals watts $\times$ time, where time is in seconds. Assume 60 A at 12 V for 3600 s. This is 2,592,000 J.

9. The number of ampere-hours is 720. Divide by 24 to determine the amperage; this is 30 A. This is a load resistor of 0.4 $\Omega$ across a 12-V battery.

10. The power is current $\times$ voltage. The current is 200 W$/12 = 16.66$ A. If the voltage were 6 V, the current would be 33.33 A.

**Section 1.23**

1. The energy in the $E$ field is $\frac{1}{2}\varepsilon E^2 V$, where $\varepsilon = 8.85 \times 10^{-12}$F/m. The volume $V$ is 10 cm$^3$, or $10^{-5}$ m$^3$. $E^2 = 10^{12}$ V$^2$/m$^2$. The result is $4.42 \times 10^{-5}$ J.

2. The energy stored is $\frac{1}{2}QV$. Therefore, $Q = 2 \times J/V$, where $J$ is the energy stored. $Q = 8.85 \times 10^{-11}$ C.

4. If the spacing is 0.11 cm, the volume changes to $1.1 \times 10^{-5}$ m$^3$. The change of energy is about 10%, or $0.44 \times 10^{-5}$ J. Dividing this difference by the change in dimension (which is 0.01 cm or $10^{-4}$ m) gives a force of 0.044 kg. This is about 0.1 lb.

**5.** The average power is energy/time. A continuous flow of 1 A is 10 J/s, or 10 W. 10 A flowing for 0.1 s is 1000 peak watts for 0.1 s, or 100 J/s. 100 A flowing for 0.01 s is 100,000 peak watts, or 1000 J/s.

**6.** The peak power is given in answer 5 above.

## Section 2.8

**1.** Capacitance $= \varepsilon_R \varepsilon_0 A/d$. The area is 0.03 m². The spacing is 0.0003 m. $A/d = 100$ m. $\varepsilon_R \varepsilon_0 = 8.854 \times 10^{-11}$. $C = 8.854 \times 10^{-9}$ F, or 0.00885 $\mu$F.

**2.** The charge equals $CV$. For $V = 10$ V, the charge is 0.0885 $\mu$C.

**3.** The $E$ field is 10 V/0.0003 m, or $3.33 \times 10^4$ V/m. $D = \varepsilon_0 \varepsilon_R E$ or $8.854 \times 10^{-12} \times 10 \times 3.33 \times 10^4 = 2.948 \times 10^{-7}$ C/m².

**4.** At the moment of closure, the full voltage appears across the resistor. This is 10 mA.

**5.** The $RC$ time constant is $1000 \times 10^{-5} = 0.01$ s.

**6.** The voltage after one time constant is 3.7 V. After two time constants it is 37% of 37%, or 1.4 V.

**7.** $RC = 2$, where $C = 10^{-5}$ F. $R = 2/10^{-5} = 200$ k$\Omega$.

**9.** The energy in the capacitor is $\frac{1}{2}CV^2$ or $\frac{1}{2}(10^{-5} \times 100) = 500$ $\mu$J. At 37% of the voltage the energy stored is only 192 $\mu$J. The energy lost is 308 $\mu$J.

## Section 2.13

**1.** $H = 3/2\pi r$, $r = 0.1$ m. $H = 4.77$ A/m.

**2.** For one turn $H = 0.1/(2\pi \times 0.05$ m$) = 0.32$ A/m. For 100 turns the $H$ field is 32 A/m.

**3.** The $H$ field is half, or 16 A/m.

**4.** The $H$ field at the center is nearly the same.

**5.** Assume that the current flows uniformly in the wire. The radius of current flow at 0.5 cm is 0.5 cm. The current flow in one-half the radius is one-fourth. The $H$ field at this depth using Ampère's law is $H = 2.5/(2\pi \times 0.005$ m$) = 79.58$ A/m. On the surface the $H$ field is $10/(2\pi \times 0.01$ m$) = 159.2$ A/m. The $H$ field is proportional to the radius. At the center the $H$ field is zero.

**Section 2.23**

1. The current increases linearly with time. $V = L$ (amperes/second)$= 2$ V. Current changes at $2/10^{-3}$, or $2 \times 10^3$ A/s. To reach 10 mA or 0.01 A, the time is $0.01/(2 \times 10^3) = 5$ $\mu$s.

2. The energy is $\frac{1}{2}LI^2$, or $0.5(10^{-3} \times 10^{-4}) = 0.05$ $\mu$J.

3. If 20 mA flows in 500 turns, this is 1 A per turn for 20 cm. This is 5 A/m.

4. The $B$ field equals $\mu_0\mu_R H$, or $4\pi \times 10^{-7} \times 20{,}000 \times 5 = 0.0314$ T, or 3140 G.

5. Flux level equals $B \times$ area, or $0.0314 \times 5 \times 10^{-4} = 15.7$ $\mu$Wb.

6. The voltage per turn is 0.04 V. This voltage is equal to the change in $B$ field per second $\times$ the area. The area is 5 cm$^2 = 5 \times 10^{-4}$ m$^2$. Therefore, $B$ changes at $0.04/(5 \times 10^{-4}) = 80$ T/s. Divide by $\mu_0\mu_R$ to get $H$. $H$ is changing at $80/(20{,}000 \times 4\pi \times 10^{-7})$ A/m per second $= 3183$ A/m per second. For 500 turns and a 20-cm path length, this is 1.27 A/s. It takes 25.4 ms to reach 20 mA. It takes 38.1 ms to reach 30 mA.

7. At 80 T/s it takes $1/160$ s to reach 0.5 T. To reach a maximum $B$ field in the opposite sense and return to zero, it takes $4/160$ s, or $1/40$ s. This is a frequency of 40 Hz.

8. The energy stored is $\frac{1}{2}HBV$, where $V$ is volume. The volume is $7.5 \times 10^{-5}$ m$^3$. $H = B/\mu_0\mu_R = 1/(20{,}000 \times 4\pi \times 10^{-7}) = 39.79$ A/m. $B = 1$ T. Energy $= 1.49 \times 10^{-3}$ J.

9. From problem 8, the $H$ in the gap is 39.79 A/m. The current per meter in 100 turns is 0.3979 A/m. In a gap of 0.1 cm$= 0.001$ m, the current required is 0.397 mA.

**Section 3.3**

1. 1 $\mu$F at 100 Hz has a reactance of $1/2\pi fC = 1.59$ k$\Omega$. This is 159 $\Omega$ at 1 kHz and 15.9 $\Omega$ at 10 kHz.

2. 1 H at 100 Hz has a reactance of $2\pi fL = 628$ $\Omega$. This 6.28 k$\Omega$ at 1 kHz and 62.8 k$\Omega$ at 10 kHz

3. The impedance at 100 Hz is $\sqrt{1590^2 + 159^2} = 1580$ $\Omega$. At 1 kHz the impedance is $\sqrt{159^2 + 159^2} = 224$ $\Omega$. At 10 kHz the impedance is $\sqrt{5.9^2 + 159^2} = 159.8$ $\Omega$.

4. The impedance at 100 Hz is $\sqrt{628^2 + 6280^2} = 6311$ $\Omega$. At 1 kHz this is $\sqrt{6280^2 + 6280^2} = 8881$ $\Omega$. At 10 kHz this is $\sqrt{62{,}800^2 + 6280^2} = 63{,}113$ $\Omega$.

5. The peak current in the resistor is 2 mA. The reactance of the capacitor is $1/2\pi fC = 7957$ $\Omega$. The peak current in the capacitor is $10/7.597 = 1.32$ mA. The current total is $\sqrt{2^2 + 1.32^2} = 2.40$ mA. The reactance is $10/0.0024 = 4166$ $\Omega$.

6. At 5 kHz the reactance of the capacitor drops by a factor of 5: 1591 $\Omega$. The peak current is 6.60 mA. The total current is $\sqrt{6.60^2 + 2^2} = 6.90$ mA. The impedance is $10/0.0069 = 1149$ $\Omega$.

7. The current in the resistance is 2 mA. The reactance of the inductor is $2\pi fL$, or $6.28 \times 2000 \times 0.02 = 25.1$ k$\Omega$. The current is $20/25.1$ mA $= 0.797$ mA. The total current is $\sqrt{2^2 + 0.797^2} = 2.15$ mA. The impedance $= 20/0.00215 = 9302$ $\Omega$.

8. Assume a frequency of 1 kHz and a voltage of 1 V. The reactance of the 1-mH inductor is 6.28 $\Omega$. The reactance of the 3-mH inductor is three times this, or 18.84 $\Omega$. The current in the first inductor for 1 V is 0.159 A. The current in the other inductor is one-third of this, or 0.053 A. The total current is 0.212 A. The reactance is $1/0.212 = 4.72$ $\Omega$. The inductance is the reactance divided by $2\pi f$, or 0.75 mH.

## Section 3.7

1. The slope is $2\pi fV = 6.28 \times 10,000 \times 10 = 628,000$ V/s. The minimum slope is zero.

2. The natural frequency is $1/2\pi\sqrt{LC}$. $L = 0.001$ H and $C = 10^{-8}$ F. $LC = 10 \times 10^{-12}$. The square root is $1.73 \times 10^{-6}$. The natural frequency is 92.0 kHz.

3. Using problem 2, the resonant frequency is a factor of 10 lower, or 9.20 kHz.

4. The reactance of the inductor is $2\pi fL$, or 57.8 $\Omega$. The current in the inductance is $10/57.8$, or 173 mA. This is the same current that flows in the capacitor but of opposite sign.

5. The phase angle is 90°. The current leads the voltage at all frequencies.

6. The current lags the voltage by 90° at all frequencies.

7. The reactance of the capacitor at 1 kHz is 159 $\Omega$. The frequency needs to be lower by the ratio $159/1000$, or 159 Hz. At this frequency the phase angle is 45°.

8. A 0.1-H inductor at 1 kHz has a reactance of 628 $\Omega$. The frequency needs to be raised by the ratio $2000/628$ for the reactance to equal the resistance. The frequency is 3.18 kHz. At this frequency the phase angle is 45°.

## Section 3.9

1. The peak value is 12.5 V. The rms value is $12.5 \times 0.7070 = 8.84$ V.

2. The peak is $118 \text{ V} \times 1.414 = 166.8$ V.

3. Assume a 1-$\Omega$ load. The power for 1 V is $\frac{1}{4}$ W (1 W for a quarter of the time). The power for 2 V is 1 W, the power for 3 V is $\frac{9}{4}$ W, and the power for 4 V is 4 W. The total power is 7.5 W. The same power would be dissipated with a steady $\sqrt{7.5}$ V. The rms value is 2.74 V.

4. The heating power is $\sqrt{6^2 + 8^2} = 10$ V.

5. The resulting voltage is $12 \times 1.414$ less $1.2 \text{ V} = 15.77$ V. If the voltage sags 1.577 V, the average value is $15.77 - \frac{1}{2} \times 1.577 = 14.98$ V.

## Section 3.20

1. The pulse lasts 10% of the time. Assume a 1-$\Omega$ load. The average power is 10 W. This is 3.16 V dc. Therefore, the rms value is 3.16 V.

2. Energy travels at about 0.3 m/ns. The 20-m round trip takes about 66 ns.

3. The current on the initial wave is 0.2 A. It takes 0.13 $\mu$s for a round trip. There are two round trips in 0.3 $\mu$s. The current triples for each round trip. At 0.3 $\mu$s the current is multiplied by 6, or 1.2 A.

4. Given 2 A at 300 Hz. The $H$ field is $I/2\pi r$ or $2/2\pi = 0.32$ A/m rms, or 0.45 A/m peak. The $B$ field peak is $\mu_0$ times $H = 0.45 \times 4\pi \times 10^{-7} = 5.65 \times 10^{-7}$ T. The voltage induced in a loop is proportional to area (0.1 m) and $2\pi f$, or 6.28 times 300. This factor is 188.4. Multiply this factor by $5.65 \times 10^{-7}$ T. The peak induced voltage is 0.106 mV.

5. The capacitor is charged to $14.14 \text{ V} - 0.6 \text{ V} = 13.54$ V. The charge is $CV$ or $1.35 \times 10^{-4}$ C. 10 mA for 10 ms is $0.01 \times 0.01 = 10^{-4}$ C. The remaining charge is $0.35 \times 10^{-4}$ C. The voltage remaining is proportional to charge, or $0.35/1.35 \times 10 \text{ V} = 3.51$ V.

6. The heating voltage is the square root of the sum of the squares, or $\sqrt{12}$ V. The power dissipated is $12/10 = 1.2$ W.

7. At 90° the voltage is $10\sqrt{2} = 14.14$ V. At 45° assume that one pointer is horizontal. The other has a horizontal component of 7.07 V and a vertical component of 7.07 V. The horizontal components add to 17.07 V. The vertical component adds by the square root rule, or $\sqrt{1707^2 + 7.07^2} = 18.48$ V.

8. The current $= C \times$ the rate of change of voltage $= 5 \times 10^{-12} \times 10 \times 10^6 = 50$ $\mu$A.

## Section 4.5

1. The area of the wire is $\pi r^2$, where $r$ is 0.05 cm= 0.0079 cm². The length is 100 cm. The resistance is 1.72 $\mu\Omega$ × 100/0.0079 $\Omega$ = 21.8 m$\Omega$.

2. The resistivity ÷ 5 is 5000 $\Omega$.

3. 10 in. = 0.4 $\mu$H. 120 in. = 48 $\mu$H.

4. 0.02 in. = 0.051 cm. The resistance = 1.72/0.051 = 33.1 $\mu\Omega$. There are 10 squares in series, or 331 $\mu\Omega$.

5. The resistance is 10/0.01 = 1000 $\mu\Omega/\square$.

## Section 4.10

1. The $H$ field = $E/377$ = 53.05 mA/m.

2. $\lambda/2\pi$ = 100 m, $\lambda$ = 628 m. The wavelength is 300 m at 1 MHz. Thus 628 m is the wavelength at 0.478 MHz.

3. The $H$ field falls off linearly in the far field. 2 A/m × 20/30 = 1.33 A/m.

4. The interface distance is $\lambda/2\pi$. $\lambda$ = 150 m. The distance = 23.9 m.

5. The interface distance at 10 MHz is $30/2\pi$ m. The wave impedance halves for a current loop or 188 $\Omega$.

6. The wave impedance doubles, or is 754 $\Omega$. $E$ increases by 1.414, or 14.14 V/m.

7. The $H$ field at the interface distance is 10/377, or 26.5 mA/m. The $H$ field is 26.5/1.414 = 18.76 mA/m.

8. The effective radiating power is 50,000 W. The $E$ field at 20 km is $\sqrt{30P}/r$. This is 61.24 mV/m. At 40 km the $E$ field is half, or 30.62 mV/m.

## Section 4.13

1. The skin depth reduces as the square root of frequency. At 1 MHz the skin depth is 0.0062 cm. At 4 MHz the skin depth is 0.0031 cm. The $E$ field that penetrates the surface is 0.1 V/m. At one skin depth the field is reduced to 37%, or 0.037 V/m.

2. At two skin depths the field strength is 37% less, or 0.014 V/m.

3. The ratio of 1 MHz to 400 Hz is 2500. The square root is 50. Therefore, the skin depth is 0.0062 × 50 = 0.31 cm.

4. The skin depth is proportional to the square root of resistivity. At 1 MHz the skin depth for copper is 0.0062 cm. The ratio of resistivity to copper is $2.83/1.72 = 1.65$. The square root of this ratio is 1.28. The skin depth is $1.28 \times 0.0062$, or 0.0079 cm. At 400 Hz the frequency ratio is 2500. The square root is 50. The skin depth is $0.0079 \times 50 = 0.40$ cm. At 60 Hz the square root ratio is 129.1. The skin depth is thus $129.1 \times 0.0079 = 1.02$ cm.

5. The skin depth decreases by the square root of the permeability ratio $\sqrt{100/1}$ and increases by the square root of the resistivity ratio, or $\sqrt{1.72/10} = 0.41$. The resulting skin depth at 1 MHz is $0.0062 \times 10 \times 0.41 = 0.025$ cm.

## Section 4.23

1. The half-wavelength is 7.5 m. The screw spacing of 10 cm is one aperture that attenuates the field by $10/750 = 0.0133$. The field is 0.133 V/m. For the ventilation holes the maximum dimension is the diagonal, or 3.2 cm. This attenuates the field by $3.2/750$, or a field strength of 0.043 V/m. The meter diagonal is 10 cm. The field from this aperture is 0.133 V/m. The field from the three sources is 0.31 V/m.

2. The half-wavelength at 100 MHz is 1.5 m. The opening of 1.5 cm attenuates the field by $100:1$. The width/depth ratio is $4:1$. The waveguide attenuation is $10^6$. The attenuation at the end of one opening is $10^8$. There are 50 openings, so the field is attenuated by a factor of $2 \times 10^6$.

3. The half-wavelength at 10 MHz is 15 m. The 2-, 3-, and 4-cm openings attenuate the field by $2/1500$, $3/1500$, and $4/1500$. The field strengths are 0.0027, 0.0041, and 0.0054 V/m. The wire screen attenuates the field by a factor of $0.005/15$. The attenuated field strength is 0.0006 V/m. The sum of the fields is 0.0125 V/m.

4. At 10 MHz the half-wavelength is 15 m. At 100 MHz the half-wavelength is 1.5 m. The coupling for 10 m at 10 MHz is $10/15 \times 2 \times 20\text{V/m} \times 0.2$ m $= 5.33$ V. At 100 MHz the WCC requires that the factor $10/15$ be removed. The result is 8 V.

## Section 4.30

1. The $B$ field must transition from $-0.1$ T to $+0.1$ T in 10 $\mu$s, or 0.02 T/$\mu$s. The area of the core is $10^{-4}$ m$^2$. The rate of change of flux is equal to 0.02 T $\times$ the core area, or $2 \times 10^{-6}$ Wb/$\mu$s $= 2$ Wb/s. This is equal to 2 V per turn. 20 V requires 10 turns.

**2.** $1/\pi\tau_r = 318$ kHz. The impedance of the $RC$ circuit is 583 $\Omega$. A 100-V signal at this frequency causes a current of 171 mA. The peak capacitor voltage is 85 V.

**3.** The starting point of 316 $\mu$V is given in Section 4.8. Now multiply by the following factors: frequency, 0.1; loops, 100; area, 1; distance, 1; voltage, 10; impedance, 30/500. This is 1.90 mV/m. The 100 loops of 2 cm$^2$ contribute 3.8 mV/m. The total is 5.7 mV/m.

**4.** The current supplied for 110 real watts is 1.06 A. The reactance of the magnetizing inductance is $2\pi fL$, or 1880 $\Omega$. The magnetizing current is 0.062 A. These currents add together on the diagonal of the current rectangle. This is 1.062 A.

**5.** The frequency of interest is $1/\pi\tau_r$. This is 3.18 MHz. The peak $H$ field at 2 cm is simply $A/2\pi r$. This is $0.1/(6.28 \times 0.02) = 0.80$ A/m. The $B$ field is equal to $\mu_0 H$, or $10^{-6}$ T. The flux in 2 cm$^2$ is $2 \times 10^{-10}$ Wb. The rate of change of flux is $2 \times 10^{-4}$Wb/$\mu$s, or 0.02 Wb/s. This is 0.02 V. The radiation at 3 m is 316 $\mu$V/m times $\sqrt{3.18}/100$, the frequency ratio; times 5, the voltage ratio; times 100, the area ratio; and times 30/50, the impedance ratio. The radiation is 17 mV/m. See Section 4.8.

## Section 5.5

**1.** 100 Hz has a period of 10 ms. 2° represents 2/360 of this time, or 55.6 $\mu$s.

**2.** The error at the input is 2/100,000 V, or 20 $\mu$V. The gain 100 makes this signal 2 mV at the output.

**3.** The 2 V is rejected by a factor of 100,000 at the output. The output error signal is 20 $\mu$V.

**4.** The common-mode rejection at 4 kHz is 10,000, or 200 $\mu$V at the output.

## Section 5.11

**1.** The energies before are $\frac{1}{2}CV^2$. This is $(\frac{1}{2} \times 10 \times 10^{-6} \times 10^2) + (\frac{1}{2} \times 10 \times 10^{-6} \times 20^2)$ J. This is 2.5 mJ. The charge is $CV$, or $(10 \times 10^{-6} \times 10) + (10 \times 10^{-6} \times 20) = 3 \times 10^{-4}$ C. The parallel capacitors are 20 $\mu$F. The voltage $= Q/C = 300$ $\mu$C/20 $\mu$V = 15 V. The new energy is $\frac{1}{2} \times 20 \times 10^{-6} \times 225 = 2.25$ mJ. The energy is lost in arcing.

**2.** 440/5 A = 88 A.

**3.** The reactive current is 2 kW/220 V = 9.09 A. The reactance is 220/9.09 = 24.2 $\Omega$. $1/2\pi fC = 24.2$ $\Omega$. $C = 109.6$ $\mu$F.

4. The reactance at 120 Hz is $1/2\pi fC = 13.3\ \Omega$. The rms voltage is 0.707 V. The rms current is 53.2 mA.

5. 2 A and 20 m$\Omega$ is 0.04 V. There are two connections, so the lost voltage is 0.08 V.

## Section 5.13

1. The excess gain is 1000. The output impedance is reduced to 0.1 $\Omega$.

2. 5 $\Omega$ at 10 kHz looks like $5/2\pi f = 80\ \mu$H.

3. The input impedance at 1 kHz is 1 M$\Omega$. The capacitance is $C = 1/2\pi f \times 10^6$. This is 159 pF.

## Section 5.17

1. The time of storage is $\frac{1}{2}$ cycle or 10 $\mu$s. The current flow is 100 mA. The charge is $0.1 \times 10^{-5}$ C. $C = Q/V$, or $10^{-6}/0.1 = 10\ \mu$F.

2. The frequency of interest is $1/\pi\tau_r = 3.2$ MHz. The impedance of the circuit is 200 $\Omega$. The radiation is 316 $\mu$V/m times 3 for area, divided by 200/30 for impedance, multiplied by 20 for voltage, and divided by 1000 for frequency. The answer is 2.84 $\mu$V/m.

3. The 200-$\Omega$ load reflects to the primary 200/9 $\Omega$. The 400-$\Omega$ load reflects to the primary 400/4 $\Omega$. The power is $12^2/22.22 + 12^2/100 = 7.92$ W. The capacitances reflect by factors of 9 and 4. The total capacitance is 1300 pF.

## Section 5.20

1. The energy in the inductance is $\frac{1}{2}LI^2$. This is $0.25 \times 0.024^2 = 144\ \mu$J. The voltage across 400 pF to store this energy is $\frac{1}{2}CV^2 = 144\ \mu$J. $V = 848$ V. The energy in the arc is 144 $\mu$J.

2. If the $R$ value is 100 $\Omega$, the current is multiplied by 10, and the energy stored is 100 times as great. The voltage rises by 10, to 8480 V.

3. The time constant is $L/R$, or $0.5/1000 = 0.5$ ms. Two time constants = 1 ms.

4. If $R = 100\ \Omega$, $L/R = 5$ ms. Two time constants = 10 ms.

## Section 6.9

1. The capacitance of a parallel-plate capacitor is $\varepsilon_0\varepsilon_R A/d$, where $\varepsilon_0 = 8.854 \times 10^{-12}$ F/m. The area is 0.02 m² $\times d = 2.54 \times 10^{-4}$ m. $C = 697 \times \varepsilon_R$ pF. If $\varepsilon_R = 7.1$, the capacitance = 5000 pF.

2. The wave travels the length of the line in 20 ns. The current that flows in the capacitor is the short-circuit current. 10 V and 100 Ω is a current source of 100 mA. The charge that flows in 40 ns is $0.1 \times 40 \times 10^{-9}$ C. The voltage in a 10-μF capacitor is $Q/C$, or 0.4 mV.

3. The voltage drop is 1.0 V. The charge in 1 ms is 0.02 A$\times 10^{-3} = 2 \times 10^{-5}$ C. The capacitance is $Q/V$, or 20 μF.

## Section 6.15

1. The current flow in the capacitor is $C$ times the rate of change of the voltage. This is $15 \times 10^{-12} \times 10^9$ V/s = 15 mA. This is 1.5 V in 100 Ω.

2. The radiation using the standard model is 316 μV. The frequency of interest is $1/\pi\tau_R = 31.8$ MHz. Area: multiply by 57. Frequency: divide by 9.9. Impedance: multiply by 0.6. Voltage: multiply by 10. Distance: divide by 3.33. The result is 3278 μV/m.

3. If the adjacent conductor is 0.05 in. away, the field is reduced by the ratio of areas, or 364 μV/m.

4. The voltage coupled to the cable is 0.02 V/m. This is 0.04 V in 2 m. Half of the energy goes in each direction. At the unterminated end, the voltage doubles. The result is 0.04 V.

5. The half-wavelength at 20 MHz is 7.5 m. The area is 0.03 m². The voltage is 10(area/half-wavelength) = 0.225 V.

6. The area of each loop is 0.8 cm². The frequency of interest is 10.6 MHz. The radiation from the standard model is 316 μV/m. Number: multiply by 125. Area: multiply by 0.8. Voltage: multiply by 5. Frequency: divide by 89. Impedance: divide by 3.3. Distance: multiply by 1. The result is 537 μV/m.

7. Number: multiply by 500. Area: multiply by 0.3. Voltage: multiply by 5. Frequency: divide by 89. Distance: divide by 3.3. The result is 806 μV/m.

## Section 6.28

1. The area of the loop is 0.01 m². The peak voltage is 0.282 V. The maximum rate of change of flux is 0.282 Wb/s. The $B$ field is changing at $0.282/0.01 = 28.2$ T/s. The $H$ field is equal to $B/\mu_0$. The $H$ field is changing at $2.25 \times 10^7$ A/m per second. At 2 MHz the $H$-field peak is 1.79 A/m. This is 1.26 A/m rms.

**2.** The current level is 5 A at 300 MHz. The $H$ field at 10 cm is $A/2\pi r =$ 7.96 A/m. For a plane wave the $E$ field is 377 H = 3000V/m. The aperture dimension is 0.56 cm. The attenuation is 75/0.56. The field is 22.4 V/m.

**3.** Using problem 2, the $E$ field is 3000 V/m. If the aperture length is 2 cm, the field that is coupled is $2/75 \times 3000 = 80$V/m. For an area of 2 cm$^2$, the voltage coupled is 2.13 V. This can damage a circuit.

**4.** The filter must attenuate the signal by a factor of 10. If $R = 100 \ \Omega$, then the reactance of the capacitor should be about 10 $\Omega$. At 300 MHz, this is 53 pF.

## Section 6.30

**1.** The $H$ field 1 ft from the girder is $I/2\pi r$, where $r$ is 0.33 m. $H =$ 26,500 A/m. The $B$ field is $\mu_0 H$ or $4\pi \times 10^{-7} \times 26,500$ T. The area A is 200 cm$^2$ or 0.02 m$^2$. The flux is BA or $0.66 \times 10^{-3}$ Wb. The voltage is the rate of change of flux, which is $2\pi f$Wb. The frequency of interest is 640 kHz. Therefore, $2\pi f$Wb = 2,676 V.

**2.** The reactance at 640 kHz is 75.4 $\Omega$. For 20,000 A, the voltage drop is $1.5 \times 10^6$ V. The breakdown voltage for 6 in. is 300,000 V. Lightning will jump through the concrete.

# APPENDIX II
# Glossary of Common Terms

| | |
|---|---|
| Alternating current (ac) | Current that reverses direction at regular intervals. Usually sinusoidally. |
| Ampere (A) | Unit of current. |
| Analog | Continuous representation of a physical process. |
| Aperture | Opening in a conductor that allows electromagnetic fields to penetrate. |
| *B* field | Magnetic induction; in teslas. |
| Balanced | Having the same characteristics on two lines. |
| Bobbin | Device for holding coils of wire. |
| Capacitance | Charge stored per unit voltage. |
| Capacitor, *C* | Component for storing electric field energy. |
| Characteristic impedance | Impedance that terminates a transmission line so that there are no reflections. |
| Charge, *Q* | Presence or absence of electrons. |
| Charge amplifier | Circuit that converts a changing charge to a changing voltage signal. |
| Closed-loop gain | Gain when feedback is in place. |
| Coax | Shielded conductor used in high-frequency transmission. |
| Coil | Usually, turns of copper wire. |
| Common | Zero-voltage conductor in a power supply; sometimes called a *grounded conductor*. |
| Common-mode signal | Average voltage or current impressed on a group of conductors. |
| Conductance | Reciprocal of resistance; in siemens. |
| Conductivity | Reciprocal of resistivity; the property of a conductor to allow current flow. |
| Conductor | Material that allows charges to move under the influence of an electric field. |
| Core | Usually, transformer iron. |

| | |
|---|---|
| Coulomb (C) | Unit of charge. |
| Current | Flow of charge, $Q/t$. |
| $D$ field | Electric field resulting from charges. |
| Direct current (dc) | Steady fixed voltage or current. |
| Delta connection | The three legs of the power source connected to form an equilateral triangle in a three-phase transformer. |
| Dielectric | Insulator used between conductors in capacitor construction. |
| Dielectric constant | Ratio of the $E$ field in space to the $E$ field in a dielectric. |
| Differential | Operating on the difference in potential. |
| Digital | Representation of commands and values in terms of a sequence of binary values. |
| Diode | Component that restricts current flow in one direction. |
| Distribution transformer | Transformer that distributes power at required voltages for a facility. |
| $E$ field | Electric field intensity at a point in space; in volts per meter. |
| Earth | Conducting surface of our planet. |
| Equipment | Hardware, instruments, conduit, receptacles, motors, racks; usually, conductive housings. |
| Equipment ground | Conductor that surrounds or is near any electrical wiring; conduit, outlet boxes, panels. |
| Equipment grounding conductor | 1. Safety or green wire, conduit. 2. Receptacle housing, motor housing, rack frame. |
| Equipment grounding electrode system | System of interconnected conductors in a facility that make up all equipment grounds and building steel. |
| Equipotential surface | Any surface where no work is done in moving a unit charge. |
| Farad (F) | Unit of capacitance. |
| Feedback (electrical) | Circuit arrangement where the output signal is fed back to the input to control the voltage or current gain. |
| Feedback factor | Gain in excess of the closed-loop gain. |
| Ferrite | Magnetic material used in transformers. |
| Filter (electrical) | Group of components that attenuates or selects bands of frequencies. |
| Flux | Electric or magnetic field. |
| Flux (quantity) | Field crossing an area; field intensity × area. |

| | |
|---|---|
| Flux density | Field intensity at a point in space; volts per meter or amperes per meter. |
| Frequency | Repetition rate; cycles per second in hertz. |
| Gasket | Flexible conductor that bonds two surfaces together to close an aperture. |
| Gauss (G) | Unit of magnetic intensity; 10,000 gauss = 1 tesla. (*See B* field.) |
| Gradient | Slope; in volts per meter. |
| Green wire | Safety wire or equipment grounding conductor. |
| Ground | 1. Reference conductor, often a power supply return. 2. Earth connection. 3. To make a connection to a conducting surface, often earth. 4. Conducting surface, not necessarily earthed. |
| Ground loop | Multiple connections that allow interference currents to flow in sensitive leads. |
| Ground plane | 1. Group of interconnected conductors. 2. Sheet or grid of conducting material. |
| Grounded conductor | Current-carrying power conductor that is grounded at the service entrance. |
| Grounding electrode system | All of the metal structures in a building that could come in contact with power wiring or lightning. |
| *H* field | Magnetic field intensity at a point in space; in amperes per meter. |
| Henry (H) | Unit of inductance. |
| Hertz (Hz) | Unit of frequency; cycles per second. |
| Hysteresis | Nonlinear relationship between *B* and *H* in a magnetic material. |
| Impedance | Opposition to current flow when resistance and reactance are involved. |
| Inductance | Total magnetic flux generated per unit of current. |
| Inductor, *L* | Component that stores magnetic field energy. |
| Insulator | Material that does not allow charges to move under the presence of an *E* field. |
| Isolated ground | Equipment ground returned separately to a remote panel or service entrance. |
| Isolation transformer | Special shielded power transformer. |
| Joule (J) | Unit of work or energy. |
| Lamination | Shaped piece of magnetic material used to make the core of a commercial transformer. |
| Leakage capacitance | Mutual capacitance. |

| | |
|---|---|
| Leakage inductance | Magnetic flux that does not couple between coils in a transformer. |
| Listed | Approved for use by a national testing organization. |
| Magnetic flux | Magnetic intensity ($B$ or $H$)×crossing area; in square meters. (*See* Webers.) |
| Magnetizing inductance | Apparent inductance that provides a $B$ field in the core of a transformer. |
| Micro- ($\mu$-) | 1/1,000,000. |
| Milli- (m-) | 1/1000. |
| Monopole | Single magnetic pole (hypothetical). |
| Mutual capacitance | Ratio of charge induced on a grounded second conductor to voltage placed on a first conductor. |
| Mutual inductance | Ratio of magnetic flux coupled to a second loop from current flow in a first loop. |
| Neutral | Power conductor grounded at the service entrance in three-phase distribution. |
| Normal mode | Signal of interest. |
| Ohm ($\Omega$) | Unit of resistance, reactance, or impedance. |
| Open-loop gain | Gain before feedback is applied. |
| Permeability (relative) | Ratio of $H$ in a magnetic material to $H$ in a vacuum. |
| Permeability of free space | Ratio of $B$ to $H$ in free space; $4\pi \times 10^{-7}$. |
| Permittivity of free space | Ratio of $D$ to $E$ in free space; $8.854 \times 10^{-12}$. |
| Phase shift | Time shift measured in the number of electrical degrees for a sine wave. |
| Plane wave | Radiated signal far removed from the radiating source. |
| Rack | Steel structure for housing electronic hardware. |
| Radiation | Electromagnetic energy that leaves a circuit. |
| Reactance | Opposition to sinusoidal current flow in a capacitor or inductor; in ohms. |
| Rebars | Reinforcing steel bars within concrete. |
| Reference conductor | Point or conducting surface chosen to be the zero of potential. |
| Resistance | Opposition to current flow; in ohms. |
| Resistivity | Property of conductors to oppose current flow. |
| Resistor, $R$ | Component that resists the flow of current. |
| Separately derived power | Source of power (transformer) that has its neutral regrounded to the nearest point on the equipment grounding electrode system. |

| | |
|---|---|
| Service entrance | Power panel that receives the power leads for a facility or residence. |
| Shield | Usually, a conducting surface that intercepts fields. |
| Siemens (S) | Unit of conductance. |
| Skin effect | Current restricted to flow on conducting surfaces at high frequencies. |
| Solenoid | Coil of wire used to form a magnetic field; the geometry of most small inductors. |
| Tesla (T) | Unit of magnetic intensity. |
| Thermocouple | Junction of dissimilar metals; a voltage is generated when the junction is heated. |
| Three-phase power | Method of transporting power on three conductors where the voltage peaks on each conductor 120 electrical degrees apart. |
| Time constant | Length of time required for a simple transition to within 37% of final value. |
| Toroid | Doughnut-shaped structure. |
| Transmission line | Parallel conductors that support the flow of electromagnetic energy. |
| Ungrounded conductor | Power conductor at the power voltage in single-phase power. |
| Volt (V) | Unit of electrical potential difference. |
| Voltage, $V$ | Work required to move a unit charge between two points in an electromagnetic field. |
| Watt (W) | Unit of power; work done at the rate of 1 J/s. |
| Waveguide | Cylindrical tube used to transport wave energy without a center conductor. |
| Weber (Wb) | Unit of magnetic flux; $B$ field $\times$ area in square meters. |
| Wye connection | Three-phase power; the three legs of the power source connect to one common point (neutral). |
| Zener diode | Component that conducts above a specified voltage. |

# APPENDIX III
# Abbreviations

The abbreviations used in this text are IEEE-sanctioned. In a few instances an abbreviation might have several meanings. The number of abbreviations that are used in this book may make reading difficult for some readers. This list is provided to help readers not familiar with this style of text.

| | |
|---|---|
| A | Ampere |
| A/m | Amperes per meter ($H$ field) |
| ac | Alternating current |
| $B$ | Magnetic induction field intensity in teslas (T) |
| $C$ | Capacitance or Capacitor |
| C | Coulomb (electric charge) |
| $C_{12}$ | Mutual capacitance |
| cm | Centimeter |
| $cm^2$ | Centimeter squared (area) |
| cmr | Common-mode rejection |
| CRT | Cathode ray tube |
| d | Length, Diameter |
| D | Displacement field |
| dc | Direct current |
| DUT | Device under test |
| $E$ | Electric field intensity (volts per meter); Energy |
| ESD | Electrostatic discharge |
| f | Frequency, Force |
| F | Farad |
| FAA | Federal Aviation Administration |
| FCC | Federal Communications Commission |
| ft | Foot |
| ft-lb | Foot-pound |
| GHz | Gigahertz ($10^9$ hertz) |
| $H$ | Magnetic field intensity (amperes per meter) |
| Hz | Hertz (cycles per second) |
| $I$ | Current |
| $IC$ | Integrated circuit |

| | |
|---|---|
| IEEE | Institute of Electronic and Electrical Engineers |
| in | Inch |
| J | Joule |
| k | Kilo (1000), Factor used in rating power transformers |
| kg | Kilogram (1000 grams) |
| kHz | Kilohertz (1000 cycles per second) |
| km | Kilometer |
| $k\Omega$ | Kilohm (1000 ohms) |
| l | Length |
| $L$ | Inductor or inductance |
| $L_{12}$ | Mutual inductance |
| $L/R$ | Time constant (inductance) |
| Lb | Pound |
| LISN | Line Impedance Simulation Network |
| m | Meter; Milli |
| $m^2$ | Meters squared (area) |
| mA | Milliampere (0.001 ampere) |
| mm | Millimeter |
| mV | Millivolt |
| mW | Milliwatt |
| $m\Omega$ | Milliohm |
| M | Mega ($10^6$) |
| MHz | Megahertz |
| MOV | Metal oxide varister |
| $M\Omega$ | Megohm ($10^6$ ohms) |
| n | Nano ($10^{-9}$); Number |
| ns | Nanosecond |
| NEC | National Electrical Code |
| p | Pico ($10^{-12}$) |
| pF | Picofarad |
| pC | Picocoulomb |
| $Q$ | Charge |
| $r$ | Radius |
| $R$ | Resistor or resistance |
| RC | Resistor-Capacitor; Time constant |
| RL | Resistor-Inductor |
| RLC | Resistor-Inductor-Capacitor |
| rms | Root mean square |
| s | Second |
| S | Siemen (conductance) |
| t | Time in seconds |
| T | Tesla ($B$ field) |
| V | Volt; Volume |
| V/m | Volts per meter ($E$ field) |

| | |
|---|---|
| W | Watt |
| Wb | Weber |
| WCC | Worst case calculation |
| $X_C$ | Capacitive reactance |
| $X_L$ | Inductive reactance |
| Z | Impedance |
| $\varepsilon_0$ | Permittivity of free space |
| $\varepsilon_R$ | Relative permittivity (dielectric constant) |
| $\lambda$ | Wavelength |
| $\mu$ | Micro ($10^{-6}$); Permeability |
| $\mu_0$ | Permeability of free space |
| $\mu_R$ | Relative permeability |
| $\mu A$ | Microampere |
| $\mu F$ | Microfarad |
| $\mu H$ | Microhenry |
| $\mu s$ | Microsecond |
| $\mu V$ | Microvolt |
| $\pi$ | 3.1416 (pi) |
| $\rho$ | Resistivity |
| $\Omega$ | Ohm |
| $\Omega/\square$ | Ohms per square |

# INDEX